An award-winning writer on science, nature, and the environment for *Time* magazine, *National Geographic*, and other publications, EUGENE LINDEN is the author of six books, including the landmark study *Apes, Men, and Language*, *New York Times* "Notable Book" *Silent Partners*, and *The Future in Plain Sight: Nine Clues to the Coming Instability*. He has consulted for the U.S. State Department, the United Nations Development Program, and is a widely traveled speaker and lecturer. Mr. Linden has donated a portion of his royalties from *The Parrot's Lament* to the Humane Society of the United States and to Traffic, a branch of the World Wildlife Fund dedicated to stopping the trade in endangered species.

THE PARROT'S LAMENT

And Other True Tales of Animal Intrigue, Intelligence, and Ingenuity

EUGENE LINDEN

A PLUME BOOK

PLUME
Published by the Penguin Group
Penguin Putnam Inc., 375 Hudson Street, New York, New York 10014, U.S.A.
Penguin Books Ltd, 27 Wrights Lane, London W8 5TZ, England
Penguin Books Australia Ltd, Ringwood, Victoria, Australia
Penguin Books Canada Ltd, 10 Alcorn Avenue, Toronto, Ontario, Canada M4V 3B2
Penguin Books (N.Z.) Ltd, 182–190 Wairau Road, Auckland 10, New Zealand

Penguin Books Ltd, Registered Offices: Harmondsworth, Middlesex, England

Published by Plume, a member of Penguin Putnam Inc. Previously published in a Dutton edition.

First Plume Printing, August 2000
10 9 8 7 6 5 4 3 2

 REGISTERED TRADEMARK—MARCA REGISTRADA

The Library of Congress has catalogued the Dutton edition as follows:
Linden, Eugene.
   The parrot's lament: and other true tales of animal intrigue, intelligence, and ingenuity / Eugene Linden.
     p. cm.
   Includes bibliographical references.
   ISBN 0-525-94476-1 (hc.)
      0-452-28068-0 (pbk.)
   1. Animal intelligence. 2. Animal behavior. I. Title.
QL785.L725 1999
591.5—dc21
                                        99-23424
                                          CIP

Printed in the United States of America
Original hardcover design by Eve L. Kirch

BOOKS ARE AVAILABLE AT QUANTITY DISCOUNTS WHEN USED TO PROMOTE PRODUCTS OR SERVICES. FOR INFORMATION PLEASE WRITE TO PREMIUM MARKETING DIVISION, PENGUIN PUTNAM INC., 375 HUDSON STREET, NEW YORK, NEW YORK 10014.

For Mary

ACKNOWLEDGMENTS

MY greatest debt is to my parents and most particularly my mother, Gloria Linden, who from my earliest years brought to life for me the individuality and personality of the animals we encountered in the course of the day. This environment of rampant anthropomorphism left a lasting imprint on me, manifested in high school when I chose as my subject for a paper on Greek mythology the minor demigod Melampus, whose singular gift was that he could talk to the animals.

I've been writing about animals and whether they could talk or think ever since. In the twenty-eight years since I began work on *Apes, Men, and Language*, I've had the benefit of conversations and guidance from hundreds of scientists, trainers, and keepers, many of whom I have acknowledged and cited in previous books and articles.

Still, I would like to single out a few scientists and animal people who showed particular willingness to help me understand some of the intelligence, virtues, and vices of various animals. These include Beth Armstrong, Gail Laule, Maureen Fredrickson, Charlene Jendry, Gigi Ogilvie, Rick Glassey, Helen Shewman, and Geoffrey Creswell.

The administrators and keepers at the Woodland Park Zoo, the San Diego Zoo, the Bronx Zoo, the Columbus Zoo, the Tulsa Zoo, and ZooAtlanta, were particularly helpful and cooperative, as were the officials of the American Zoo and Aquarium Association, the American Veterinary Medicine Association, the International Society for Behavioral Ecology, and the Society for Interative and Comparative Biology. Many thanks to Leon Levy and Shelby White, who provided a perfect place for me to hole up when I was making the final revisions to the manuscript. Thanks also to Katherine and Aubrey Buxton, who brought to my attention the extraordinary story of Harriet the leopard.

In the course of this book I revisited a number of scientific and philosophical issues that I have written about previously, and I would like to thank James Gould, Richard Byrne, Sally Boysen, Lyn Miles, Donald Griffin, Thomas Sherratt, Karen Pryor, William Conway, Gail Patricelli, Daniel Dennett, Steve Nowicki, Marion East, and Heribert Hofer for their insights and wisdom.

The book also benefited from the competence and professionalism of the people at Dutton. I'd like to thank Claire Ferraro for supporting the book every step of the way, Karen Lotz for her superb editing and boundless intelligence, and Karen's assistant, Amy Wick, for fielding problems, questions, and requests way beyond the call of duty. I must thank David Bjerklie for making sure I policed the text for unjustifiable assumptions (a difficult job in a book whose very premise is unjustifiable in the view of some scientists). I would like to thank Cary Ryan for giving me the benefit of her keen eye for lapses in internal logic. And, as always, I would like to thank Esther Newberg for providing me and the book with the best possible representation.

Finally, if a little sentimentality can be permitted, I have to thank the many pets who have come and gone in my life, starting with Cinders, my childhood cat; Zephyr, the heroic Maine Coon cat; Murghatroyd, Philo, and Junior, our present Bengals, and Caruso, an ancient and wise stray cat in our neighborhood who somehow has made friends with all the people on our street even while beating up all of our pets. All have told me something about what it is like to be an animal. Even Bully, my giant bullfrog who, when I was ten, ate all the other frogs in his enclosure, was telling me something, although I'm still not sure what particular message he was trying to get across.

CONTENTS

PREFACE

Animals appear in so many guises in various aspects of my writing that if a psychiatrist ever got me into analysis, he or she might reasonably assume that apes, dolphins, elephants, birds, felines, and canines have some totemistic importance in my worldview—that in our fellow creatures I see embodiments of every aspect of the human situation. The shrink would be right, of course. That's exactly how I view the world. How we treat animals in captivity reflects on our view of ourselves and our humanity. How we treat wild nature gives us insight into whether we are going astray in our dealings with the rest of the natural order.

The course of my writing has taken me to the far corners of the globe, and animals pop up even when my purpose is to write about entirely different matters. For instance, I went to Antarctica in the fall of 1996 with the ostensible mission of writing about the relationship of the frigid continent to the global climate system. Once there, however, I found one of the more compelling ways to convey the delicate balance of climate and life in this extreme environment was to look at the impact of an early breakup of the McMurdo Sound sea ice on the population of emperor penguins. That Antarctic summer, the sea ice broke up two weeks early, forcing fledglings to begin their wanderings through the Southern Ocean before they were ready to embark, probably dooming most to an early death.

More often when I've headed off to some remote place, I'm specifically looking for animals. I've been to places in central Africa where the animals have never before encountered humans, and I've been to places in southeast Asia where scientists are discovering large animals previously unknown to science. I've visited almost every major research station for studying the great apes in the wild. I've traveled to rain forests in Borneo, India, and various parts of the Amazon to report on the loss of biodiversity as we humans crowd countless species into extinction.

The creatures with whom we share the planet are interesting in their own right as objects of study and observation, but they can also tell us who we are, and where we came from. Let me give you an example. I was originally drawn to write about animal intelligence by simple curiosity. I first heard about the experimenters who were attempting to teach language to chimps in 1970, and I thought, "Gee, isn't language the ability that is used to distinguish between animal and human? What would it mean if animals had some capacity to use human language?" What indeed?

After a stint in Vietnam as an investigative journalist in 1971, I got around to looking into this question. Over the years I have written two books, *Apes, Men, and Language* and *Silent Partners*, coauthored another, and contributed several articles on various aspects of the primate studies and the controversies they inspired to a wide variety of magazines, including *Time*, the *National Geographic*, the *New York Times Magazine*, and the *Atlantic Monthly*. In the course of these various writings, I learned a bit about language, a bit about animals, and a lot about the nature of science—and scientists.

Today, I'm as interested in animal intelligence as I have ever been. I still believe—hypothetical psychiatrists take note—

that animals can tell us who we are and where our abilities came from. But I have no desire to write another book about studies of animal language and animal intelligence.

I think I've found a more interesting, if unscientific, way to get at these questions. For those readers who have no stake in the issues that so heavily weigh upon the question of animal intelligence, I suggest you skip right to the introduction, because the stories and anecdotes that this book offers, while revealing, are above all fun. Those readers curious about why I chose my particular approach in this book (yes! I've found a new way to write about animals!) might want to continue with this preface.

First, a few cautionary words are in order about the dragons and sea monsters that lurk just beyond the horizon when anybody sets sail on an exploration of animal intelligence. The seemingly simple question of whether apes or other animals can learn language goes to the very heart of our concepts of human nature and our alleged special place in the natural order. The debate summons emotions and inspires vitriol like few other issues in science. At a symposium on animal consciousness at the 1999 meeting of the Society for Integrated and Comparative Biology in Denver, Colorado, Harriet Ritvo, a historian of science based at MIT, described the issue of animal consciousness as "contentious," by which she meant an issue that cannot be argued without reference to ideology. This also beautifully describes the debate over animal language.

It is a sad comment on the behavioral sciences that even in the course of thirty years, a period during which the chimp named Washoe, the first chimp to be taught sign language, has gone from childhood to old age and a new generation of behavioral scientists has entered the field; a period during which genetics has moved from its first crude attempts to recombine the building blocks of life to actually cloning mammals; a period during which computers have progressed from slow and costly number crunching to replicating knowledge and pervading every aspect of life; indeed, a period during which there has been tremendous progress in the comparative study of other higher mental abilities of animals—that during this period of great progress in knowledge on so many fronts, the debate about whether animals might acquire some aspects of human language has advanced on a time scale that might best be described as geologic. It has been easier to end the Cold War and defeat communism than it has been to find agreement among scientists about what an animal means when it uses human language.

There is now widespread agreement that apes and some other higher mammals can understand and use words in semantically appropriate ways, but when any scientists move beyond that to assert evidence that an animal has generated or responded to a complex sentence, they suffer death by a thousand pricks as critics question their methods, their data, their controls against cheating, their rigor, their honesty, and sometimes, it seems, their patriotism. At one conference devoted to the issue some years back, there were even dark murmurings of scientific fraud, murmurings that were hastily retracted when the murmurers were challenged to cite specifics by Nicholas Wade (then of *Science*, more recently of the *New York Times*) at the postsymposium press conference.

In short, while there have been a great number of studies, a consensus about whether human language is different in kind or only in degree from the communicative abilities of animals is not much closer than it was in 1968. Since that time, a number of scientists hoping to skirt the endless debates about animal language have moved into the study of animal cognition and consciousness. There has been some scientific progress in this new tack (which I will discuss later), but here again scientists find themselves in endless "contentious" debates about what they mean by consciousness and awareness. As the philosopher Daniel Dennett put it, "Overshadowing otherwise straightforward studies is the huge problem of how you talk about the experiments."

The problem is whether subjective experiences can be discussed objectively. At the Denver symposium, John Stadden, a theoretical psychologist from Duke University, described succinctly why strict behaviorists believe that the study of consciousness is not a scientific endeavor. Since ultimately all knowledge of mental events is by its nature private, goes this argument, and science is concerned with data that is public, then the study of internal mental states is not a proper subject for science. While this position has a certain consistency, it amounts to saying that one of the single most important aspects of human nature is unknowable. As Donald Griffin, a pioneer in promoting the study of animal thinking, has remarked, if consciousness is important to us and it exists in other creatures, then it is probably important to them. He also said that the study of nature is difficult enough without going at it with one hand tied behind one's back and a foot in a bucket.

In fact, there is a way to explore the mental world, including the communicative abilities, of animals without getting trapped in this colloquy from hell. It's unscientific, which means that at the end of the exercise, no matter how persuasive the material, one cannot use the stories and examples as proof of anything. I suspect, however, that a lot of people can live with that, because I also suspect that there are a lot of people like me who have watched animals in both nature and captivity solve problems and use their wits (yes, wits!) to get what they want. Unmindful of the debate about their abilities, animals use what abilities they have. What they do is interesting. The debate about what they do might once have been interesting, but it certainly isn't anymore.

What I have in mind, then, is simply to examine stories about the ways in which various animals react to humans, not in the context of scientific study, but in the course of their daily lives. Most zookeepers, for instance, have no stake in the question of whether there is continuity between animal communication and human language, and many have only informal scientific training. But they have not waited for the solons of linguistics and the behavioral sciences to certify that some animals can understand complex utterances. In fact, many would not care if a thousand behaviorists got together and issued a definitive statement that what appears to be understanding is no more than the animal responding to various nonverbal cues that keepers make when they talk to their charges. If telling an animal to meet you around the corner gets an appropriate response, a keeper is going to continue to issue this verbal command because it is the simplest way to communicate.

Despite the unending repetitions of the debate about animal language abilities, these decades of criticism of various experiments have been useful in some respects. For one thing, critics have explored every conceivable way in which a simple response to body language might be confused with a higher mental ability. Indeed, up to a point the critics' skepticism is justified. One need only consider the case of Clever Hans.

In the early part of the century, this horse dazzled crowds in Europe with his apparent ability to do simple arithmetic at the command of his trainer. His owner would ask, "How much is two and three?" and the horse would begin tapping his hoof on the ground, stopping with uncanny accuracy at the right number. An intrigued Dutch psychologist named Oskar Pfungst convinced Hans's owner to try an experiment in which Hans performed his feats while blindfolded. Unfortunately, when unable to see his trainer, Clever Hans turned into a dunce. It turned out that rather than counting, Clever Hans had been closely watching his owner, who would visibly relax when the horse got to the right answer.

What looked to the owner and all observers like arithmetic was merely the natural sensitivity of a large prey animal to the body language of the key figure in its life.

More recently, a British psychologist named Richard Byrne did a small experiment to investigate the seemingly uncanny communication that takes place between humans, border collies, and the sheep they herd. On a trip to the Little Rann of Kutch in Gujurat, India, Byrne spent some time hanging out with a herd of sheep observing the interactions between the shepherds, the sheep, and their herding dogs. According to Byrne, the dogs played a role in the herd, but they were more oriented toward exterior threats. He also noticed that the shepherds moved and herded the sheep with the same whistling sounds that would appear to be commands for the dogs. It turns out that not only the dogs but the sheep understood what the shepherds were communicating, and the sheep were responding to the humans. (As a humorous aside, Byrne notes that in New Zealand, sheep of different types can learn to stand next to the appropriate sign, lettered in English, for their breed, leading some observers to conclude that they knew how to read.)

Such cautionary stories are useful for ground-truthing a particularly compelling anecdote since it is easy to get carried away by the mental agility of an animal that is doing little more than paying close attention to the body language of its owner. On the other hand, while Hans may not have been "intelligent" enough to perform sums, he was still pretty clever to figure out a way to give his owner an answer that would yield him a treat.

Those brave souls who have continued to investigate the comparative study of language and intelligence also have provided an enormous amount of data about various aspects of higher mental abilities that will help put the stories in the following chapters in context. Louis Herman of the University of Hawaii has shown that dolphins respond appropriately to changes in word order, suggesting that they understand how the placement of words in a sentence can change and even reverse its meaning. Sue Savage-Rumbaugh has explored similar problems with bonobos (also known as pygmy chimpanzees) in Atlanta, and Ron Schusterman has probed this aspect of language with sea lions and even elephant seals at the Long Marine Laboratory in Santa Cruz.

Other scientists have deconstructed various forms of deception and tried to determine whether animals can be aware of another creature's mental state (one clue to consciousness). All of this work helps us gauge how much credibility we should attach to some of the stories that follow, and in the course of this book, I will refer to various experiments that might help put an anecdote in context.

But above all, this book is about stories—not studies—of animal intelligence. I hope you enjoy them.

INTRODUCTION

Max and Patty are two of the elephants at the Bronx Zoo in New York City. To enliven their daily routine, Chris Wilgenkamp and the other keepers have introduced a number of games. One is a form of pachyderm "hide-and-seek" in which the humans will secretly cache a favorite object like a tambourine or a tire, and then will let the elephants try to find it using their sense of smell. When the huge mammals are getting "warmer," Chris will blow a whistle; when they are getting "colder," Chris does nothing. According to Chris, the elephants love the game, especially the cues. "They get all excited when they are getting close," he says, "and when they find what we've hidden, they flap their ears." Apart from keeping the elephants active, such games can provide a fun way to deliver medicines and vitamins that have been mixed in and hidden with the treats.

All in all, this little vignette of life in the zoo provides a nice example of how humans can enrich the lives of captive animals while dealing with health and nutritional issues at the same time. But to leave the story at this point would miss something else that is going on in the elephant enclosure. It turns out that the keepers are not the only ones inventing ways to make zoo life more interesting. Even as the humans introduce fun activities, the elephants are developing their own games as well.

At the end of each outdoor session, the keepers will put rewards inside the gates of the night cages to entice the elephants to come back in. Once they are inside, the keepers shut the gates and secure the exhibit for the night. Max and Patty seem to have figured out that the keepers will only shut the gates when both elephants are inside. So when the keepers lay out the enticing bedtime treats, one will go in while the other lurks just outside the gates. Then the first elephant will come back out while the other goes in to get his or her share.

The motivation seems pretty simple—the game provides a chance to prolong their time outside and maybe thumb their trunks at the keepers. But think about it: the game also requires that one elephant trust the other not to eat all the treats. What is going on here? Are the animals communicating and planning, or did they arrive at this apparent strategy by blind trial and error? Jot down your own interpretation of this little game; we will come back to it later.

We encounter animals continually and ubiquitously in our everyday lives. These encounters, however, are usually so routine (at least for us), whether they involve petting the dog or cat in the morning or going to the zoo, that for most of us, animals are little more than simple components of the landscape—"Did you see that beautiful bird at the feeder?" "My cat Molecule killed another mouse today." We might pause and chuckle for a moment when we hear about a pet that does something "smart" like learning to flush a toilet, or when we watch a television special about an elephant that likes to paint, but most people spend very little time wondering about what, if anything, is going on in an animal's head when it interacts with people. It's easy to forget that every other creature on the planet is, like us, engaged in its own active struggle to survive, and that for an enormous number of animals, a large part of life involves figuring out how to deal with the dominant species on the planet— humankind.

There are people, however, who are, by necessity, very aware that animals are not just a passive part of the land-scape. These are the veterinarians, researchers, and zoo-keepers who deal with animals on a daily basis; people like Chris Wilgenkamp. Some are scientists, some dedicated ama-teurs. Most are not studying animal intelligence, but they en-counter it, and the lack of it, every day. Get a bunch of these people together and inevitably they will start telling stories about how their charges either tried or succeeded in out-smarting, beguiling, or otherwise astonishing the humans in their lives.

Such tales are the material for this book. These stories are about attempts by animals to deceive or manipulate their keepers or each other, stories about games, stories of under-standing and trust across the vast gulf that separates differ-ent species, stories of animal heroism, and, especially if the keepers have had a few drinks, stories about escape.

Such was the situation at the 1998 annual meeting of the American Zoo and Aquarium Association in Tulsa, Okla-homa. Sitting around the table in the cocktail lounge of the convention center Doubletree Inn were keepers from Dis-ney's Animal Kingdom in Florida, from ZooAtlanta, from the Columbus Zoo, and other keepers from a number of other zoos who stopped by to say hello. Almost everybody at the table was smoking cigarettes. In demeanor the keepers reminded me of cops: they have high-stress, demanding, and sometimes dangerous jobs; they are more comfortable with each other than with outsiders. But as the evening wore on, they began to relax and decided that it was OK to tell me some of their stories. When the conversation turned to animal escape attempts, Charlene Jendry, a veteran gorilla keeper from Columbus, said, "Tell him about the wire." Another keeper smiled, nodded, and said, "Oh yeah, Fu Manchu."

Fu Manchu was a male orangutan who had lived at a number of zoos. His most memorable escapes took place at the Omaha Zoo in Nebraska over thirty years ago. Not once but twice keepers arrived at the outdoor exhibit to discover Fu Manchu as well as a female orangutan and their three offspring up in trees outside the exhibit area. Both times the keepers opened the doors of the display area, cajoled the family back inside, and diligently double-checked all the locks. A few days later, however, the keepers again arrived to find Fu Manchu and family enjoying the morning sun on the roof of the exhibit.

Only after the keepers set up a watch did they solve the mystery. On nice warm autumn days when the orangutans were allowed into the outdoor area, Fu Manchu would head for the moat. In the wall of the moat was a door that led into the furnace room. On the other side of the room were stairs and another door that led to freedom. Drawing on his great strength (keepers have watched orangutans playfully twist steel-belted radial tires into pretzels), Fu Manchu pulled the door back from its frame. Taking a piece of wire from his cheek, he then tripped the latch, much the way a thief might slip a credit card between a door and its frame.

Once the keepers took the wire away from him, Fu Manchu's escapes ended. But plenty of his successors have carried on this distinguished orangutan tradition. Within the zoo community, orangutans are legendary for their escape attempts. Papillon and other virtuoso breakout artists have nothing on orangutans in terms of the patience, improvisation, and sheer bravado they bring to bear in their escape attempts. Not only are orangutans by far the most imaginative escape artists, but their facility with human tools continually amazes those humans who deal with them. Keepers are fond of quoting a remark made by Ben Beck, a zoologist who was one of the designers of Think Tank, an ambitious program to explore the cognitive abilities of orangutans at Washington's National Zoo. He once noted that if you give a screwdriver to a chimpanzee, it will try to use the tool for every-thing except its intended purpose. Give one to a gorilla, and it will first rear back in horror—"Oh my God, it's going to hurt me"—then try to eat it, and ultimately forget about it. Give it to an orangutan, however, and the ape will first hide it and then, once you have gone, use it to dismantle the cage.

Make no mistake, other apes and indeed any number of zoo animals will try and succeed in escape attempts (some of which I will describe later), but orangutans have no peers when it comes to the frequency and ingenuity of their at-tempts. Zoo designers take every type of precautionary mea-sures when it comes to trying to anticipate when and how an orangutan might breach a barrier. They hire rock climbers to determine whether there are hidden handholds they might not have anticipated. They put "hot wires" with a mild elec-trical current at the top of a wall. No matter, in one Texas zoo, an orang grabbed clumps of grass to make an insulating mitt, then climbed over the hot wire without harm.

Of course, humans could easily design cages from which no escape is remotely possible, but that is not the point of a zoo, particularly since the current trend is to try to create as natural a display as possible without jeopardizing the wel-fare of the animals or those who come to see them. So zookeepers and orangutans are trapped in the pongid equiva-lent of an endless arms race in which designers try to come up with exhibits that seem natural and will still keep the ani-mals inside, while the orangutans probe every weak point or hidden flaw the builders might not have noticed.

There is one major difference between those daring hu-man prisoners who have escaped from captivity and their orangutan equivalents: motivation. Both ape and human may feel that they have been unjustly imprisoned, but many keepers believe that a number of complex motivations drive orangs to continually test their keepers, ranging from a simple desire to outsmart and turn the tables on the humans who control their lives to curiosity about the strange world outside the walls and moats that contain them. (Rather than head for the hills, in most cases, ape escapees hang around. Orangs and gorillas might explore the grounds, or, as in the case of one orangutan at the National Zoo in Washington, take a piece of fried chicken from a traumatized tourist and eat it.)

Once outside, except for those animals raised among humans, most escapees are terrified. For better or worse, the zoo offers three squares a day, companionship, and security. Most escaped animals have not learned the survival skills to get by in the wild by themselves. Even if they could get by, more often than not escaped animals would find themselves in places with completely different climates and surroundings from those of their native habitat. At some level, most captive animals recognize that the zoo is where they must live.

This has been brought home on many occasions. When high winds destroyed the netting that made up the outer layer of the New York Zoological Society's huge aviary in the 1960s, zookeepers moved those birds who remained and left the netting open. According to William Conway, president of the society, almost 75 percent of the escaped birds returned within two days. When another storm collapsed an aviary in 1995, the escaped birds hung around but could not get back in because the structure had collapsed in on itself.

And then there was the case of the orangutan found eating fried chicken, an example that beautifully underscores the complex motivations that underlie escape attempts. Rob Shumaker, who with Ben Beck helped design Think Tank at Washington's National Zoo, which he now coordinates, recalls that a new keeper had mistakenly left some big green barrels in the outdoor enclosure with the animals. Rob had no idea that anything was amiss until word got back to him that Bonnie, one of the female orangs, had been spotted eating fried chicken and drinking orange juice not far from a concession stand. As Rob reconstructed the event, she had simply stacked the barrels and exited, followed by other orangutans ready for adventure. The chicken and juice came from a cooler she had taken from a terrified tourist.

Later, after the animals had been led safely back into the enclosure, a tourist came up to Rob and asked, "Did you get 'em all back in?" Rob, a little perturbed and very perplexed that the orangutans had been out for some time before anyone contacted him, was in no mood for casual conversation. "You saw them outside?" he asked testily.

"Yes," the man answered, a little taken aback by Rob's tone.

"Didn't it occur to you to tell someone?" Rob continued.

"No," replied the visitor. "We didn't think it was anything unusual since they'd been going in and out all morning."

Clearly, for these orangutans, this escape was not about getting back to Indonesia. For every story of escape, there are many stories of animals that would rather remain in their cages, however bleak. Some years ago in the Congo I came upon what for me was the most poignant story illustrating the conflicted feelings that accompany captivity. This was at the Pointe-Noire Zoo, a dilapidated relic of the colonial era with ancient cages. Among the few animals remaining was a small colony of chimpanzees. One of these, an old female appropriately named La Vieille, lived in a small concrete cage with a barred gate as its front. It was hard to imagine how the poor animal could live in the cage for more than a month, much less the years that she had spent there.

Much harder to fathom, however, was why she stayed inside. The latch on the gate had broken many years earlier, and the gate swung freely on its rusty hinges. The old female could have left the cage at any time, yet she didn't, apparently preferring the impoverished security of her home to the terrors of the outside world.

Escape is but one of the many reactions that animals have to captivity that can reveal some degree of cleverness. At the opposite end of the range are stories that illuminate the extraordinary degree of trust that, despite all the inconvenience of captivity, the animals feel for their human masters. There are of course some dunderheaded and mean-spirited keepers in the zoo community, as in any profession, but the great majority choose the vocation because they love animals. All the zookeepers I spoke with talked about their charges in some respect the way parents talk about their children—with amusement at their pranks and foibles, and genuine pride in their feats. Eventually conversation among animal keepers turns from tales of mischief to stories of insight, trust, and understanding.

Charlene Jendry, for instance, tells of one extraordinary moment with a gorilla named Brigette at the Columbus Zoo. In 1986 Brigitte gave birth to Fossey, the first baby gorilla reared by its mother at the zoo (named for Dian Fossey, the larger-than-life figure who was murdered in Karisoke, Rwanda, the field station she established to study mountain gorillas). One day, Charlene walked past the cage and noticed that Fossey was nursing sloppily. Without thinking, she said to Brigette, "That kid's got milk all over his face. Bring him over here and I'll wipe it off." Without hesitation, Brigette brought the baby over and pushed his face up to the bars so that Charlene could wipe it clean.

Did she know what Charlene was saying? An innocent enough question—but as noted in the preface, the issue of how much of human language animals can understand has been a subject of intense debate in the behavioral sciences during the more than three decades since R. Allen and Beatrix Gardner first taught a chimp named Washoe to make some of the signs used by the deaf in American Sign Language and another scientist named David Premack taught a chimp named Sarah to communicate using plastic tokens. What then can we make of Brigette?

Brigette never had any formal training in understanding English. Did she bring the baby over and put his face up to the bars to be wiped on some impulse of her own, unrelated to Charlene's request, or did she vaguely understand some fragment of what Charlene said, like "Bring it over here," a phrase that is often used when the gorillas get their hands on some forbidden object? Or did she understand that Charlene was saying that the baby had a messy face and that if Brigette brought him up to the bars of the cage, Charlene would clean it?

If this was the only example of a gorilla responding appropriately to a casual comment, it might be dismissed. But Charlene has many more examples that suggest the gorillas pay close attention to their human masters. She remembers that one day Molly, a twelve-year-old female, was very sick. Charlene and another keeper named Adele Absi decided to try to take the gorilla's temperature by using a paper-strip thermometer they had purchased at a drugstore. Charlene, who had never used these strips before, experimented by putting the strip on her forehead. Molly was watching, her foot right next to the wire mesh, while Charlene and Adele discussed how they were going to take her temperature. The two keepers decided to put the strip on Molly's foot in order to get a reading, but they weren't sure whether this would work. It didn't matter. With no prompting, Molly took the strip from her foot and placed it on her forehead. Once a reading appeared, Charlene and Adele asked Molly to bring the strip over, and Molly took the paper from her forehead and brought it to the keepers. It is possible that Molly was mimicking something she had seen Charlene do, but her other actions suggest that she understood in some way that the strip might help her keepers help her.

Well, maybe Charlene Jendry is unreliable, romantic, or prone to overinterpretation. If so, there is an epidemic of mass hysteria among keepers around the country, because I have been flooded with stories of apes' eavesdropping on human conversation or responding appropriately to complex statements. As I probe further, it turns out that most keepers make a practice of choosing their words carefully around gorillas, chimps, and orangutans because the animals are always listening. At the Woodland Park Zoo, for instance, keepers use code words when discussing one particular female orangutan because she becomes paranoid when she hears her name mentioned. When Kyle Burks was at the Atlanta zoo, he would say to Ivan, an adult male gorilla, "Go around the corner and we'll meet you," just like he might say it to a friend.

Ivan was initially raised by humans, which means he spent his formative years in a language-rich environment. Most keepers will agree that the more time an animal spends with humans, the better it seems to understand spoken language. At ZooAtlanta, the keepers have noticed a difference between the English competence of orangutans like Allen and Hanti, who occasionally respond to spoken English phrases appropriately, probably cued by the context and tone of voice, and Chantek, an orangutan who for fifteen years was part of a long-term language experiment undertaken by Lyn Miles at the University of Tennessee in Chattanooga. "If we tell Allen or Hanti to 'go get the paper,' they may eventually show up with it," says Laura Mayo, a keeper at the zoo, "but if we give the same instruction to Chantek, he will search his cage, find the paper, and bring it to us."

Before coming to ZooAtlanta, Chantek spent several years at the Yerkes Regional Primate Center, where his keepers, hoping to reintegrate him with other orangutans, refused to communicate with him in American Sign Language, the language that had served as his primary means of communication with humans for the prior fourteen years. Although Chantek forgot a number of signs during the several years at Yerkes, he still uses many signs appropriately to this day. Indeed, his new keepers, like Laura Mayo, say that they are learning sign language from the orangutan. They will point to a tomato, for instance, and ask Chantek the word for tomato, and he will make the appropriate sign.

Chantek has even invented signs to tell the keepers what

he wants. When he feels that it is time for a meal, he will look at a keeper and tap his wrist as though tapping a watch in the universal gesture of telling someone that it is getting late. He likes his food scattered on the floor of his cage when it is delivered, and when a keeper approaches with a bucket of food, he makes a scattering gesture.

What one man can do, another man can do, goes the saying, and it is no less true of apes and other animals. Within any species there will be brighter individuals and dim bulbs, but if one animal demonstrates understanding of a sentence or complex proposition, you can bet that others within the species are capable of that same understanding.

Thus, while scientists debate whether apes can understand and use language, many keepers assume that apes understand what they are saying and actually count on that understanding to make their jobs easier. At the Bronx Zoo, Chris Wilgenkamp says that two of the elephants, Happy and Grumpy, have learned some English words like *tambourine* (a favorite plaything) and *tree* (another toy), and that when he wants the elephants to get something in one of the games they play, he will simply say, "Go find it."

Let's now reconsider that opening anecdote about the game two Bronx Zoo elephants invented to prolong their time outside. It turns out that there are many other examples of animals coordinating their response to humans in ways that suggest some form of sophisticated animal-to-animal communication. I witnessed one such example when I visited Lou Herman at the Kewalo Basin Marine Mammal Laboratory in Hawaii. In this particular case, he was probing the dolphin's ability to cooperate in response to a human command. The dolphin subjects of this experiment were Akeakemai and Phoenix, two bottlenose dolphins who lived in a tank at the laboratory. Somewhat compensating for the relative confinement of the tank was the near constant attention they received from graduate students and volunteers, who engaged them in a variety of stimulating activities.

In this case, two instructors stood on a platform on the edge of the tank. Holding a finger high in the air, the instructors caught the dolphins' attention and issued a command in a series of gestures. First, the instructors tapped two fingers of each hand together, forming the gesture that has been taught to mean "tandem." Then, they threw their arms up in an expansive gesture that meant "creative." Akeakemai and Phoenix had just been told, "Do something creative together."

The dolphins broke away from the instructors and submerged in the six-foot-deep water. There they circled until they began to swim in tandem. Once in sync, the dolphins leapt into the air and simultaneously spit out jets of water before plunging back into the water.

Even in this controlled setting, critics could find fault. Perhaps one dolphin is just very closely following the lead of the other. But if so, how would it know to take in water before jumping? Nor do the dolphins perform the same routine every time trainers ask for a "tandem creative." On other occasions they might backpedal and end the routine with a wave of their tail flukes or do synchronized back flips. The instructors never know what they are going to do.

Scientists justifiably revere Occam's razor, which holds that until proven otherwise, scientists should hew to the simplest adequate explanation for an event. In 1894 a pioneering psychologist, Lloyd Morgan, came up with an analogous formulation for understanding mental abilities that came to be called Morgan's canon. It holds that "in no case may we interpret an action as the outcome of the exercise of a higher psychological faculty if it can be interpreted as the outcome of the one which stands lower in psychological scale."

In the case of the tandem creative, some complicated bit of imitation is certainly possible, but it may not be the most economical explanation. (Looking at this question of scientific parsimony another way, if consciousness is understood to be the product of evolutionary forces, then the most reasonable position to take is that it or its precursors exist in many other creatures.) The simplest explanation may be that the dolphins listen to the command, communicate with each other and agree on a routine, and then do it. In any event, Herman's work makes it easier to consider the possibility that other highly social mammals like Max and Patty, the Bronx Zoo elephants, might be capable of planning and coordinating a response to a human signal. Yet in considering how these animals coordinate their responses, keep in mind that the possibility they are cuing each other in some manner does not guarantee that this is what they are doing.

Another type of ground-truthing has to do with the cognitive capacities of the animals themselves. Animals like apes, dolphins, elephants, and parrots live in complex and fluid social groups—a lifestyle that can foster the development of a number of skills that also serve them well in interactions with humans. The bigger the brain, the better equipped the animal is to pull off a mental stunt like an imaginative bit of treachery. One reason for this is that the ability to pull off artful bits of treachery in order to improve one's status in large social groups may improve an animal's prospects and may partly account for increases in brain size in the evolution of apes, humans, and some other highly social mammals. In any event, stories involving "smarter" animals will have disproportionate representation in this book.

That said, I've learned to keep an open mind. Some years ago Donald Griffin told me a story about miscommunication among bees that suggests that even insects might be aware when they have been given a bum steer. As first described by Karl von Frisch in 1946, when a bee scout finds a good source of pollen, it will return to the hive and do a type of waggling dance that communicates by its nature and vigor the whereabouts and amounts of the discovered riches. In this particular experiment, conducted by ecologist James

Gould on Carnegie Lake in Princeton, New Jersey, the scientist brought some flowers and bees out to the middle of the lake in a rowboat, while another group of bees was brought to a feeder closer to the shore. Once released, the foragers returned to the hive and did their dance, telling the other bees the direction of the flowers. Evidently, the bees back on dry land reacted with disbelief to the suggestion conveyed by the bee dance that there was a source of pollen in the middle of the lake; almost no bees showed up at the rowboat. On the other hand, large numbers of bees came calling for pollen when a rowboat was close to shore.

When I first spoke to Griffin about this story, he was only putting forth an amusing anecdote. In the intervening years, Griffin's position on this has become more radical. He now makes the case that even some very small-brained creatures might have some smidgen of consciousness. In practical terms, he argues, nature might find it more efficient to endow life-forms with a bit of awareness rather than attempting to hardwire every animal for every conceivable eventuality. At the symposium on animal consciousness in Denver, he told a skeptical but respectful roomful of scientists that because consciousness provided a means to test out ideas in the mind without the risks of the real world (as argued by the philosopher Karl Popper), it was possible that even invertebrates had some primitive means of "mentally" testing outcomes.

The question of awareness will come up again and again in the course of this book. Glimpses of animal awareness can surface in many categories of behavior. Some of the most moving involve moments of empathy, understanding, and trust. One of my favorites took place in northern India and was documented on film by Survival Anglia. It involved a leopard and one of that nation's most celebrated and eccentric conservationists. Over the years Billy Arjan Singh has taken into his house a number of orphaned big cats. Some years back he was sharing quarters with both a tiger cub and a leopard cub. The late German nature filmmaker Dieter Plage captured some extraordinary moments on film, including a hilarious walk that Singh took on his property with the tiger, the young leopard (whom Singh named Harriet), and his dog. The dog, who clearly had taken serious leave of her senses, at one point playfully attacked the leopard. Keeping her claws carefully retracted, Harriet responded in the spirit of the game, boxing the dog's face with a lightning series of jabs.

After Harriet reached maturity, Singh returned her to the wild, a forest preserve that lay just across the river from his compound. She adapted successfully, mated, and had two cubs of her own. Singh witnessed these events from a distance and many of them were documented by Plage on film.

Then floods came to the area, jeopardizing Harriet's den. She decided to take the babies to the one place she knew that offered safety. Carrying the two cubs in her jaws one by one, she swam across the river and moved them from her den to the "high ground" of Singh's kitchen on the second floor of his house. Respecting the prerogatives of a full-grown leopard and the eccentric priorities of Billy Singh, the kitchen staff promptly moved their cooking operations to temporary quarters.

When the floodwaters receded, Harriet went on a reconnaissance mission and inspected the den. Deciding that her wild home was once again safe, she swam back to get her cubs. The current, however, was still dangerous. So, instead of swimming across the river with a cub in her mouth as she had earlier, she hopped into the prow of Singh's dugout (in which she had ridden many times before), and then looked back to him as if to say, "What are you waiting for?" Singh took the hint and ferried the leopard and her baby across the river. The story shows extraordinary trust extending across the species barrier. It is also hard to resist the idea that Harriet "knew" how to take advantage of both the interspecies relationship and human technology (the boat) to protect her cubs.

There are other examples of animals turning to humans when their offspring are in jeopardy. Such stories are noteworthy because to allow a human to minister to a sick or failing infant, a mother has to suppress her most profound protective instincts. One of these incidents occurred at the National Zoo in Washington, and Rob Shumaker says it was a watershed moment in his appreciation of animal awareness. It happened about ten years ago, when Bonnie's infant son, Kiko, became seriously ill.

The infant was being raised by the mother in the orangutan complex, and the keepers faced a dilemma. The baby needed an injection, which meant separating him from his mother. They could tranquilize the mother to get at the baby, but they were reluctant to do so because it would be traumatic for both mother and baby.

Before resorting to that measure, Rob asked the veterinarians whether he could try something. He went up to the bars with a syringe in his pocket. Taking it out, he showed it to Bonnie, who was holding the baby in the back of the cage, and said, "Bonnie, I need to give the baby this." Bonnie had undergone injections herself and knew what a syringe was and that shots hurt. But, it appears, she also knew that what was in the syringe might help heal her baby.

First she stared at Rob. Then, holding the baby, she brought him up to the bars and turned him around so that Rob could have access to him. He gave the baby the shot in his thigh, and even though the infant screamed bloody murder, Bonnie held him still until Rob had finished giving the injection. Only then did she take him away from the bars and console him. "She trusted me," says Rob, "but what I found so extraordinary was that she appeared to understand the situation and willingly participated in it."

There are also stories about animals that show devotion that seems indistinguishable from what we know as love.

Writing in the *National Geographic*, Douglas Chadwick told of one such story that again took place in India. Asian elephants are used in logging operations and develop very strong bonds with their mahouts, who are the animals' trainers, providers, and companions. Chadwick writes about one mahout who had a tendency to drink himself into unconsciousness. On those occasions, the man's elephant gently would use his tusks to pick him up, carry him to his home, and then carefully deposit him at his front door. Not long after the mahout passed away, the elephant died as well. "Maybe," wrote Chadwick, "he went to find his mahout."

Sometimes an animal's affection for humans spills over into romance, a phenomenon that can give new meaning to the word *crush*. Once when I was visiting an orangutan rehabilitation center in Borneo, I had to deal with the advances of an adolescent female who would drop onto me from the roof of my hut when I exited and wrap her arms around me in an amazingly powerful hug. (Having witnessed the flirting of a female gorilla, I know an ape come-on when I see it.) At Karisoke, the research station where generations of scientists have studied mountain gorillas, one young woman from the University of Michigan found it hard to keep a proper scientific distance from a rambunctious young male gorilla who would do cartwheels in front of her and then grab her arm or shoulder and drag her off in an apparent display of gorilla/caveman affection.

Animals have also from time to time risen to what we would consider the heroic. Groups like the Delta Society in Seattle, which promotes the use of companion animals, have collected countless stories of cats waking their owners when the house is on fire, or dogs dragging babies to safety, but some documented actions reveal extraordinary awareness and sensitivity. When, on August 19, 1996, a toddler fell eighteen feet into the moat surrounding the gorilla exhibit at the Brookfield Zoo in Chicago, an eight-year-old female lowland gorilla named Binti Tua kept the other gorillas away from the unconscious child, then gently picked him up and delivered him to the keepers who had rushed to the service gate of the exhibit.

And then there is grief. One story involves two star-crossed orcas, named Orky and Corky, who found themselves in an inadequate tank at a now defunct park called Marineland in Palos Verdes, California. A number of critics say that the facility was not large enough for the orcas to thrive, and the animals suffered a number of ailments. The staff, however, was completely devoted to their welfare, and the orcas had been comfortable enough to attempt to breed. Unfortunately, Corky never produced a baby that lived past infancy. Gail Laule and Tim Desmond, who both now consult with various zoos on animal behavior as employees of Active Environments Inc., worked with Orky and Corky as keepers and trainers during their years at Marineland. According to Laule and Desmond, one of Corky's tragic pregnancies drove Orky to try to hurt himself in his grief.

In the 1970s it was difficult to monitor orca pregnancies. The keepers did not know that Corky was pregnant with her first baby until she gave birth. During another pregnancy, Orky would swim up beside Corky and put his forehead against her belly, presumably using his fantastic acoustic imaging abilities to give Corky a sonogram. At one point, when everything looked reasonably normal with the mother to the keepers, Orky gave his mate his own examination, running his rostrum up and down her side four or five times, in just the manner a doctor might run sonogram equipment over a pregnant woman's stomach. Whatever he detected upset him greatly. Orky went over to the wall and slammed his head against the side of the tank in an outburst of emotion. Two hours later, Corky aborted. Gail and others who worked there assume that Orky realized the baby was dead as soon as he put his head to Corky's belly.

These, then, are the types of interactions I will explore in this book. Escape figures high among the priorities of some captive animals, but I will also offer tales of deception, greed, manipulation, and revenge as well as heroism, loyalty, trust, and love. If animals have some degree of awareness, after all, they are likely to share human vices as well as virtues. I will also offer stories in which animals put skills they learned from humans to work in their dealings with other animals. One of the more hilarious examples of this involves a parrot, and the story merits retelling here because it also opens up the tricky question of anthropomorphism.

Sally Blanchard is a parrot shrink, although she balks at the term and prefers to be referred to as an animal be-haviorist. Operating out of a small home in Alameda, California, she consults with parrot owners with problem birds. At first the idea sounds narcissistic, self-indulgent, and, well, Californian. In a state that has become emblematic of wretched excess in all areas of self-improvement, why not put your parrot on the couch if it screams all day or obsessively plucks out its feathers?

It turns out, however, that Sally Blanchard and other parrot behavioral specialists fulfill a basic need. Unlike dogs and cats, domesticated for thousands of years, parrots are wild animals, not at all adapted to human company. Their needs and neuroses are not nearly as well understood as those of other pets, and these smart, highly social animals are capable of making life a living hell if an owner inadvertently presses a hot button. Thus, parrot owners from all over the country call Sally to consult. When they do, they also tell her stories about things their birds have done. Some are just too good to be true, and one of the best involved an African Grey in Arkansas.

One thing that parrots and their kin can do better than any other animal is say human words, and African Greys are among the most loquacious and gregarious of the entire family. I've seen more than ten thousand African Greys alight at the edges of a clearing for the evening, and their calls and squawking raise an incredible din. They are extremely social birds, and Jimbo, the Grey being raised in Arkansas, was no exception.

Jimbo's human mother, Bev Llewellyn, lives on a farm near Strickler, Arkansas, where she and her husband, Bill, raise Shetland sheepdogs. She has a strong relationship with her parents, who live in a mobile home on the property, not far from their daughter's house. Bev would regularly take the parrot over to her parents' for dinner. Her mother would play "peekaboo" with the parrot (the game seems to be a favorite among the species), and the bird learned to call her owner's parents Poppa and Peekaboo. Indeed, when Bev and Bill would start to get ready to go over to the trailer for dinner, Jimbo would say, "Can we go to Poppa's and Peekaboo's for dinner, okay? Come on!" In all probability unaware of the internecine implications, the parrot also learned to say, "We're having tacos and chicken," among a number of other phrases.

On one particular occasion, the bird was sitting in her cage near a window that looked onto the patio (long after they had named the parrot, the Llewellyns discovered that Jimbo is a she). Looking out the window, Jimbo saw a wild roadrunner and said to Bev, "Momma, look! A bird!" Then Jimbo turned to the wild bird just outside the window and said, "Hi, bird. Are you hungry? Do you want to go to Poppa and Peekaboo's for dinner. Peekaboo cooks chicken and corn."

What can we make of this? It's tempting to believe that the parrot was generously extending a dinner invitation to a wild bird, but that would be anthropomorphizing—imposing human motives on other creatures. This is a scientific no-no, and such an assumption is probably not warranted in the example cited above. It is unlikely that the parrot would really care that much about the plight of a strange bird of another species. On the other hand, a few words about anthropomorphism are in order before getting into the stories themselves.

Anthropomorphism comes from a natural tendency to see human behavior and motives in the actions and attributes of other creatures. The octopus looks smart because above its very appealing eyes is a bulbous headlike structure that appears to house a big brain. It doesn't—that's the octopus's body. Similarly, people go the extra mile for dolphins in part because of their noble foreheads, their grace, and their permanent smiles. Behind that forehead, however, is echolocation gear, not brains (the gray matter is located farther back), and that smile was produced by evolution to support a feeding strategy and not to express joie de vivre. If dolphins had evolved to attack their prey fish from above rather than below, they might have been cursed with a permanent frown. This is not to say that dolphins and octopuses aren't smart—they are (and the marine mammal is smarter than the mollusk)— or that dolphins are not noble and playful. But it is important to keep in mind that we cannot judge an animal's intelligence or nature by its looks.

This is one of the lessons of our ambivalent relationship with chimpanzees. There is a lot of evidence that they are very smart, but it is their misfortune that they look like something that failed to become a human being. In the wild, chimps make tools and organize themselves for such noble endeavors as warfare and cooperative hunting (these also, of course, are familiar human pursuits). In captivity, they have learned and used words in symbolic ways, another ability previously thought to be exclusively human. But chimps are also intense and aggressive, and will not hesitate to bite people when stressed or wronged. Female chimps come into heat with huge swellings which look great to male chimps but vulgar to unenlightened humans.

Chimps are an unwelcome reminder of our chthonic roots and not the kind of creature most people are going to give the benefit of the doubt. Harriet Ritvo notes that many Victorians, already disturbed by Linnaeus's classification of apes and humans as similar, were repelled by Darwin's assertion that they had a common ancestor. This unease continues to this day. While dolphins have legions of guardian angels, chimpanzees are our poor relations who find their kinship · acknowledged only when it becomes useful: for instance, when they are used as surrogates for humans in medical testing in the search for the cure for such diseases as AIDS and hepatitis B.

If we must be careful about making judgments based on looks, we also must not leap to conclusions about behavior—even in a close relative like the chimp. When people see a grinning chimpanzee, for instance, they typically assume that the chimp is happy. In chimps, however, a grin typically signifies fear of the anger of a superior. (Come to think of it, a grin often signifies the same thing for humans in the presence of their bosses).

Despite these caveats, anthropomorphism does have its place. As hard-core behaviorists argue, objectivity is impossible when it comes to the study of higher mental abilities. The only way we can appreciate consciousness in another animal is to assume that it bears some relationship to human consciousness.

This is a controversial idea. Indeed, at the Denver symposium, Daniel Dennett spent a good deal of time arguing that the question "What's it like to be a chimpanzee?" is freighted with assumptions and not as innocent as it sounds. Dennett says that we should not assume that consciousness in other creatures is any more like human consciousness than animal communication systems are like human language. He cited a 1974 paper by Thomas Nagel in which he argued that there is no way that he or anyone else could imagine what it would be like to be a bat, whose worldview is dominated by echolocation.

In a neat reply, Donald Griffin said that he had been in touch with Nagel while he was writing that paper and had tried to persuade him to alter his conclusions. Griffin argued that blind people develop abilities somewhat similar to echolocation, perhaps offering a tiny window into how a bat or dolphin relates to surrounding spaces.

The point is that we certainly cannot learn everything about animal consciousness by considering our own feelings and thoughts, and indeed we risk being misled, but we can learn something. Any man who has been in combat or even wrestled or boxed competitively knows the profound feelings that hand-to-hand struggle summon even in the controlled situation of a match. Chimps and gorillas also wrestle; indeed, I've seen chimps use takedowns and throws that schoolboys practice for hours. It is not a stretch to assume that the fears and anxieties and occasional sense of triumph that this ancient sport evokes predate our species by millions of years, and are shared in some measure by our closest relatives.

When we think of animal consciousness, our only available starting point is our own experience. So let's begin by imagining that animal consciousness is a restricted form of human consciousness—e.g., Oskar Heinroth's definition of animals as "very emotional people with very little ability to reason"—or even posit consciousness as some collective pool from which all creatures drink, including humans. Thus anthropomorphism in the form of introspection as opposed to sentimentality may have its place in understanding what Max and Patty were up to, what the parrot in Arkansas meant, and why orangutans spend so much time trying to escape.

I'm not advocating that we go overboard—I would be very hesitant to use human feelings as a means of understanding the mind of a velociraptor, for example—but with some awareness of the many pitfalls, and keeping in mind what science has learned about those aspects of animal consciousness that can be probed through ingenious studies, we can try to imagine what it's like to be a chimpanzee, and perhaps even get a small glimpse of what it's like to be a dolphin.

These, then, are some of the approaches and assumptions that will guide this decidedly unscientific venture into the minds of other animals. The premise of this undertaking is that if we look with an open mind at the various reactions of animals to captivity, if we look at the actions they initiate in trying to outwit us, deceive us, cooperate with us, or help us, they will tell us a little about what it is like to be an orca, a tiger, or an elephant, or maybe even a bee. And once we have observed carefully, we can cautiously use our own experience to try to understand what they are telling us.

So let's see what they have to say.

THE WOLF WHO
MADE FRIENDS
WITH A GOAT

Games and Humor

G AMES are serious business. It was a game invented by an orangutan that first convinced Terry Maple, the director of ZooAtlanta and the president of the American Zoological Association, that other animals might possess some measure of intelligence. This was in 1972, when he was a graduate student doing research at the Sacramento Zoo in California.

Dr. Maple is a very large man, perhaps large enough to capture the interest of a female orangutan. Maple makes no bones about his size. Once, while proudly noting that the Atlanta Zoo had reduced the weight of a newly acquired orangutan from over 400 pounds to a trim 245, he wryly added, "Unfortunately, the zoo director has not had similar success." In any event, whether through his size or through his equally large personality, Maple certainly caught the attention of one particular orang at the Sacramento Zoo.

Maple would go into the cage area before the zoo opened so that he could observe the animals in relative solitude. As he was watching one morning, a female orangutan emerged into the outdoor area with a big rag in her mouth. Taking the rag out of her mouth, she squeezed and rolled it into a ball and then tossed it in the air a couple of times. Then she looked at Terry, and then back at the rag. Dr. Maple recalls that this back-and-forth went on for a moment, until she cocked her arm and threw the rag to the astonished graduate student. Dr. Maple did what anybody would do: he rolled up the rag and threw it back to her. After a moment, she tried to throw it back to him, but the rag landed in a spot where neither could get at it, and the game ended.

In the wild, play and games are part of an animal's education. Predators play at stalking and hunting, juvenile males of many species practice fighting and bluffing skills that will later have very serious applications, and so on. Play is also integral to our higher mental abilities. The British are fond of repeating Sir William Frasier's comment that "the battle of Waterloo was won on the playing fields of Eton." In science, the line between play and intellectual creativity is so blurred that one activity often melds seamlessly into another.

Mathematicians often use card games as the starting point for exercises in probability. In fact, Edward Thorpe, the gambling prodigy who invented a way of beating casinos at black jack, had a parallel career as a mathematician. A number of thinkers have postulated that play is the highest expression of our humanity.

While games may be essential preparation for life, there are constraints on how much time can be devoted to play. In the wild, where the margins of survival are relatively thin, a species that spends a lot of its time in frivolous pursuits is not likely to be around for very long. As is the case with so many other higher abilities, this trade-off bears on the role of play in the life of various animals. We can say that animals are playful to the degree that the value of play outweighs its costs and enhances an animal's prospects for reproduction and survival.

Captivity changes the equation by removing most of the risks associated with play. Captivity also removes constraints on an animal's time associated with food-getting and security. If an animal has a proclivity for games and humor, it has the opportunity, if not the motivation, to pursue those interests while living life among humans. Unfortunately, the terms of zoo life can be boring for animals adapted to ranging over dozens if not thousands of square miles, and boredom can suppress an animal's natural playful instincts. Some animals respond to confinement by falling into a type of lethargy. Others respond with stereotyped neurotic behavior. Still others, however, respond to unaccustomed leisure by inventing games.

At about the time Terry Maple was playing catch with an orang, I also saw a simple but memorable example of orangutan inventiveness when I visited the Oklahoma Zoo during my research for *Apes, Men, and Language*. The keepers had just cleaned the floor of the orangutan's cage, and the ape was sitting indolently in a corner of a dry part of the enclosure. After a moment or two of looking speculatively at the wet patch, the orangutan got up and, with his hands out palms down, took a running start and slid on his feet across the slick floor to the other side of the cage, looking for all the world like a surfer on a board. I was stunned because it was such a human thing to do.

More common are games that relate in some way to an animal's natural predilections. Some animals seem to decide that if they can't get to the wild, perhaps they can simulate the wild in captivity. JoAnne Simerson, a behavioral consultant at the San Diego Zoo, says that four polar bears at the zoo have invented a game that simulates a seal hunt. Buzz, the largest bear, will position himself on a diving point over the deep end of the pool in the exhibit and pretend to be asleep with a paw hanging over the edge. Then three other bears—his brother Neal and two females, Chinook and Shikari—will in succession swim up from below and touch his paw with either their nose or a paw. Buzz responds by batting them on the head. The keepers call it the bear/seal game because Buzz uses the same stroke a wild polar would

use to stun a seal. The game ends when Buzz jumps on one
of the other bears.

Some zoo animals have been known to enlist other species
in their attempts to re-create their roles in the wild. There are
even cases of animals that would ordinarily be predator and
prey joining in games that look for all the world like a benign version of what in the wild would be a life-or-death
struggle. Also at the San Diego Zoo, Simerson cites a case in
which a Rocky Mountain timber wolf and two Cretan goats
became fast friends. The wolf and the goats have adjoining
outdoor enclosures. When let outside, they will meet on opposite sides of the fence that separates their exhibits. Then
the animals will begin racing back and forth along the fence,
the goats jumping and bouncing up and down as though
in the middle of a run for their lives. Before or afterwards, to
show his friendly intent, the wolf tries to lick the goats'
faces. Having noticed how much the animals enjoy playing
together, the keepers now will bring them out at the same
time. JoAnne says that at the end of the day, the wolf will
balk at going back into his cage until he sees that his friends
are safely locked up for the night.

Halfway around the world in Africa, I came upon an
equally unlikely friendship back in 1991 when I was writing
a story on apes and humans for the *National Geographic*
magazine. In Bujumbura, the capital of Burundi, I stopped
by the home of a gold trader who kept a female chimp
named Cheetah as a pet. While the cage he had provided for
Cheetah was primitive, the female chimp had the companionship of a gigantic Rottweiler guard dog named Simba. The
formidable dog was still a puppy at heart, and loved to play
with Cheetah. Going at each other with the energy and deci-
bel level of a soccer riot, the two animals wrestled and play-
bit each other in a scene that would be R-rated for violence
were it not that throughout the madcap mayhem Cheetah
was almost helpless with laughter. It looked for all the world
as though Simba was laughing as well.

The fact that play has some serious evolutionary purpose does not preclude games from being entertaining and enjoyable. The polar bears at the San Diego Zoo may be keeping their hunting skills sharp with some idea of getting back into the wild, or perhaps they play the bear/seal game because that's just what polar bears do. But by chasing the goats, the timber wolf gets to feel a little like a real wolf, even if he's chasing his friends, while the goats get the theme-park version of the chase without the worry of becoming a meal. All over the zoo, it seems, animals are trying to express themselves as best they can, given the obvious limitations of confinement.

Around the world, zookeepers do their part, trying to encourage these creative responses to captivity by providing diversions and toys to enrich daily life for the animals, by making exhibits as natural-feeling as possible, and by creating situations in which animals can pretend to be hunting, stalking, or whatever they have been designed to do. At the Woodland Park Zoo in Seattle, Dana Wooster, who looks after the big cats, tries to make life more interesting by making mealtime a game.

She will stuff a large leaf bag with shredded newspaper and some pieces of meat. Jessie, one of the jaguars and a first-time mother, will pounce on the bag, then gut it with her hind legs to get at her dinner. Dana says that when her cubs were young, Jessie would show them how to sort through the filler in the bag to find the meat.

Wooster has toys and games for the other cats as well. The serval cat, a high-strung nocturnal feline of about thirty-three pounds, loves to play with snake skins shed by the zoo's reptiles. He will hiss at the skin, stalk it, pounce on it, and then shred it. For the bigger cats, Wooster uses bowling balls with the finger holes filled with spices like cinnamon, allspice, baking soda, and cloves (the cougar loves catnip), or she gives them large cardboard carpet rolls to attack. The clouded leopard, an endangered cat from Asia, plays a game with Dana that is similar to the play-stalking between the wolf and the Cretan goats in San Diego. Dana will pretend she can't see the leopard, and the female predator will creep up on her and pounce. The leopard must get some satisfaction from expressing skills that have been shaped by millions of years of evolution.

The sheer reward of fun becomes clear in some of the games animals invent that involve humans and man-made objects. For instance, Karen Pryor, a well-known animal behaviorist, remembers that when she worked as a trainer at Sea Life Park in Hawaii, she would take her children to swim with the spinner dolphins at the park. One of the spinners liked to play "under the bridge," and swim between her legs when Pryor was standing in a shallow part of the pool. One day this same spinner was giving Pryor's then six-year-old son Ted a ride around the pool, letting the boy hold on to its dorsal fin. With the boy holding on, the small dolphin kept approaching Karen and then stopping just in front of her. Finally she realized that the dolphin wanted to play under the bridge with Ted on its back, and so Karen told her son to get a good grip and hold his breath the next time they made a circuit of the pool. Sure enough, this time the little dolphin did not stop, but instead swam through Karen's legs with the six-year-old attached to its fin. Karen cannot explain how the dolphin knew that this time it was safe to dive with the boy on its back.

All the great apes can laugh. While a grin on a chimp may signify fear rather than pleasure, a laugh is a laugh. Over the years I have played chase-and-tickle games with gorillas, chimps, and once with a baby orangutan. All of these games end with the animal curled up laughing hysterically (or with the animal attempting to return the favor by chasing and tickling me).

Britain's John Aspinall established a spacious orphanage in Brazzaville in the Congo for gorillas whose parents had been killed by poachers. Visiting there on several occasions, I played all manner of games with the baby gorillas and a bonobo who were there being prepared for reintroduction to the wild. The young bonobo loved to be hurled high in the air. Completely comfortable in free flight, he would laugh his head off on these erratically scheduled attempts to put him in orbit. In this case, it is reasonably safe to assume that a good chuckle is as enjoyable for a chimp or gorilla as it is for a human. This is not to say that an ape is going to rival Oscar Wilde in its badinage.

Bongo, for instance, a gorilla at the Columbus Zoo, seemed to favor sight gags to wile away the time during the twenty-five years he was on display. With humans one of his games would be to play chase as he ran along the bars of his enclosure while a keeper would run along with him outside. His favorite trick was to stop suddenly, and then laugh as the keeper whizzed past outside the bars.

Like the orangutan who caught Terry Maple's attention at the Sacramento Zoo, Bongo liked games of catch. One time he was sitting on the floor of his cage with an old-fashioned water bowl and a rag. He wet the rag and rolled it into a ball. Then he harrumphed to get the attention of Beth Armstrong, the zoo's field conservation coordinator. When she was looking, he took the balled rag, tossed it into the air, and caught it with the same hand without looking. Then he said, "Hmmph." Beth started laughing when Bongo said "Hmmph," and, liking Beth's laughter, Bongo played ball games, including throwing things at Beth, for the next six weeks.

Beth says that the key to his humor was that he was a magnificent and very dignified male. He was often very detached, and Beth says she took it as a compliment that he chose to include her in his games.

Bongo was the father of Fossey, the baby gorilla mentioned in the introduction, and he would play tricks on Brigette, Fossey's mother, as well. Beth recalls that one time Bongo, Brigette, and Fossey were all sitting on the floor of the cage. Nothing much was going on, and Bongo was idly tossing a boomer ball (a small bright yellow ball that fits in the palm of your hand). Getting an idea, he looked at Brigette, then at the ball, and then back at Brigette, says Beth, as though he was measuring the distance. Then, while Brigette was still staring off into space and with a bunch of keepers looking, Bongo threw the ball and bonked her on the head. The startled Brigette threw her head back while Bongo laughed, hugely satisfied with his prank as only an enormous male gorilla might be.

Brigette was short and rotund, even by gorilla standards. Her girth proved a handicap one time when she became stuck in a rubber tub that was in the cage. Try as she might, she couldn't get out. Instead of coming to her assistance, Bongo came over and tickled her. As she struggled, both her mate and her son collapsed in laughter.

At ZooAtlanta, Terry Maple remembers a 1978 birthday celebration for Willy B., a gorilla who was about twenty at the time. As part of the fun, the keepers organized a tug-of-war with the gorilla at one end of the rope and about twenty people pulling on the other. According to Maple, Willy B. realized that even a gorilla could not outpull twenty people, and so he dropped the rope, causing the humans to fall down laughing. Enjoying the effect, Willy B. tried variations on this game, in one case waiting until everyone's guard was down and then suddenly yanking the rope to topple the human visitors.

At the National Zoo, a male orangutan named Junior once played a practical joke on Rob Shumaker. On hot days, the keepers would spray a hose in the air so that the orangutans could play with the stream of water. After a long session of such play with Junior, Rob put away the hose, signaling that the game was over. At this point, Junior was above Rob in a chute and out of sight. As Rob walked under the edge of the chute, Junior used his long arm to push a huge pool of collected water over the edge of the chute, completely drenching Rob. Shumaker says that it was clear that Junior had been waiting for the opportunity to surprise him.

In the mid-1980s I used my daughter Gillian, who is now sixteen, as a reference point for comparison between ape and human during her early childhood when I was writing *Silent Partners*. I can report with some pride that while there were strong correlations between Gillian's early learning strategies and those of apes, she has since left apes in the dust in the development of language and reasoning (although she still may be somewhat behind the great apes in terms of neatness). Humor emerges fairly early in humans, and the vaudevillian, broad laughs enjoyed by children provide a good point of comparison for understanding humor in other animals. My daughter Sofia is now two, and given that chimps seem to exhibit a degree of awareness that normally appears in humans at age four, we might comfortably use a two-year-old's sense of humor as a starting point to calibrate the sophistication of ape humor.

A two-year-old's sense of humor is not very sophisticated, but it still reflects what psychologists refer to as metacognition, or reasoning about symbolic representations of reality. Take, for instance, a joke of Sofia's that enjoyed a long run in the winter of 1998–99 as I was writing this chapter. At first I would ask her, "Are you a rutabaga?" and, giggling madly, she would reply, "I'm *not* a rutabaga!" I would then ask, "Are you a water bug?" and she would say, "I'm *not* a water bug!" (Exclamation points and italics convey only a pale shadow of the drama a two-year-old can put into these declarations.) Then I would ask, "Are you a big girl?" and after saying "Yes!" a few times, she would introduce her own little joke. Spontaneously, she started saying, "No, I'm *not* a big girl; I'm a water bug!" Then, enjoying the response she got from both her parents and anybody else in the room, she would introduce all sorts of variations: "No, I'm not a big girl; I'm a rutabaga," or the hilarious "I'm not a rutabaga; Daddy is a rutabaga." Or, finally, "I'm not a rutabaga; the baby [her one-year-old brother, Alec] is a rutabaga. I'm a water bug!"

Sofia has no idea what a rutabaga is (neither does the author), but she does know what a bug is, and apparently she realizes that the concept of declaring herself to be something that she is not is both absurd and funny. She may not yet be able to recognize that her parents can be misinformed (a critical ability in the evolution of consciousness that many parents could live without), but she does realize that by manipulating her tiny vocabulary she can create absurd and humorous propositions.

I suppose that a reductionist would say that she needn't have any idea what she is doing beyond the realization that by changing the phrase she can get a reaction from her parents. In defense of her emergent symbolic abilities, I would say that her improvisations on the game reveal an internally consistent alteration of the logic of the game as I had set it up: She had substituted "No" where "Yes" was expected and accurate, and then made an assertion that we had earlier established as contrary to fact.

There are many examples that suggest that apes can use the human language they have been taught for humorous intent, but the issue of humor, like the issue of lying, remains beyond the capabilities of rigorous testing because it is next to impossible to establish intent—e.g., "Honestly, officer, I was just kidding when I said I was going to hijack the plane." Take, for instance, this example of stubbornness and/or humor witnessed by Barbara Hiller, who worked with Koko the gorilla for a number of years in the 1970s.

Koko, a lowland gorilla who has been raised by psychologist Penny Patterson since infancy, has been taught to use American Sign Language, the language of the deaf, as a means of communication with humans. As described in *The Education of Koko*, the then young gorilla was playing by herself, arranging a nest of white towels, when Barbara noticed that she was making the American Sign Language gesture for "red." Correcting her, Barbara said, "You know better, Koko, what color is it?" Making the sign bigger each time (a way of exaggerating in sign language), Koko continued to make the gesture for "red." Finally, grinning, she picked a minute speck of red lint off the towel and held it up in front of Barbara's face.

Another much reported example of Koko's humor was an alteration of the sign for "drink." Ordinarily, this gesture involves taking a hand with the fingers bent and the thumb extended and touching the thumb to the lips and tilting the hand upwards in imitation of the way someone would tilt up a glass or bottle. After refusing to make the drink sign following repeated requests by her trainer, Koko finally relented and made the sign. But instead of holding her thumb to her lips, she made the sign in her ear.

Both of these examples could have been simple errors, but is that really the most parsimonious explanation? One way adolescents, captive animals, and great apes assert control and independence in situations in which they are dependent is through subtle forms of defiance. Koko was all three of those things at the time of that incident.

Here is a case where common sense and Morgan's canon might lead you in opposite directions. Morgan's canon holds that Koko's answers must be deemed mistakes unless there is overwhelming evidence to the contrary. The evidence that Koko was joking was powerful, but circumstantial. Scientific discipline requires us to overrule common sense where the evidence demands (e.g., that the earth orbits the sun and not vice versa, as our senses seem to tell us), but here is a case where Morgan's canon and the standard interpretation of scientific parsimony might lead us away from understanding rather than towards it.

Nor is Koko the only animal in language experiments who may have used her response to monotonous drills to make a point. Alex, the twenty-two-year-old African Grey parrot who is the subject of Irene Pepperberg's long-term study, also expresses his irritation at being endlessly asked to perform arbitrary tasks. Sometimes Alex will respond to human requests that he identify the color of a block by asking his own question. When asked, for instance, "What color . . . ?" he will say, "You tell me [pause] what color!"

Another trick Alex seems to use is to answer incorrectly so many times that it becomes blazingly clear that his errors are purposeful. Irene says that when faced with seven choices and one correct answer, Alex will sometimes give an incorrect answer twelve times in a row, choosing different incorrect answers each time, until the investigators eventually give up. Using the statistical methods scientists employ to determine chance, this makes it very likely that Alex is doing this purposefully, since the odds that he or anyone with a one-in-seven chance of getting the right answer could give twelve incorrect answers in succession are less than one in a hundred.

Tales of subtle acts of defiance abound in captive situations. When JoAnne Simerson worked with orcas, she says, one killer whale would go up to her, take her foot gently in its mouth, suck off one of her rubber diving boots, and then give it back to her. "Animals like games like this," says Simerson, "because they can exercise some control." In both the "red" incident and the "drink" incident, Koko did everything possible to show the humans around her that her use of those two signs was intentional and not an error.

Other species besides apes that have been taught human language sometimes use it to humorous effect, even if their intent remains ambiguous. Sally Blanchard, the parrot psychologist, has a number of pet birds herself. One of them, Bongo Marie, is an African Grey parrot who does not much like the other birds that Sally keeps, and Bongo Marie holds in particular disdain an Amazon parrot named Paco.

Bongo Marie's cage sits next to Sally's dining room table. One day Bongo Marie was watching as Sally cooked a Cornish game hen. Bongo Marie slid over to the side of her cage to get a better look when Sally pulled the bird out of the oven. As Sally took out a knife to cut up her dinner, Bongo Marie threw her head in the air and said with great enthusiasm, "Oh no! Paco!" Trying not to laugh, Sally said, "That's not Paco," and then showed Bongo Marie that Paco was alive and well around the corner, saying, "See? He's right over there." Bongo Marie's response was to say "Oh no" in a very disappointed voice, and then launch into a maniacal laugh.

There is little doubt that parrots enjoy getting a reaction from people. Irene Pepperberg says that the reason that parrots will imitate the sound of a microwave oven's beep or the ringing of a telephone is that people will react and come to the object that emits these sounds. In parrot lore, one of the most widespread stories is of a lost parrot that was returned to its owners when it said, "Hello, this is 557-3245. Please leave a message after the beep." This story is so widespread that it may be a parrot urban legend, but there is no question that some owners have tried to teach parrots to say the home phone number as a way of getting the birds back should they wander.

Layne Dicker, another parrot psychologist based in Southern California, seconds this. He argues that as both a prey species and a highly social species, parrots are acutely attuned to the moods of those around them. They focus on the things that humans do with energy and emotion, which, he cautions, is why one should never make exclamations during lovemaking if a parrot is around—particularly if your mother-in-law is coming to visit the next day.

Dicker lives surrounded by parrots and says that a simple joke will set off a chain reaction throughout his bird colony. Dicker, who is also a comedy writer, will tell a joke, causing his wife, Sally, to laugh. Then Hobbes, his miniature macaw, will laugh; then his Amazon parrot, Chicken, will laugh; and then, says Dicker, after all the laughter has died down, Dusky, another of his birds, will say "Hee ... hee ... hee" very slowly, causing Layne and Sally to crack up, and the whole thing will start again.

The question of what parrots do for effect, as opposed to humor, remains ambiguous, but both in the wild and in captivity they will play games with others of their own species that seem to serve only as a means of getting a reaction. In the wild, macaws and other parrots eat clay from the side of riverbanks. The minerals in the clay help them digest some of the highly acidic fruits they eat, and observers have watched a macaw high on a bank pick off a piece of clay with its foot and drop it on the head of a bird munching below.

They play similar pranks in captivity. According to Dicker, his birds will take small objects and, from the safety of their cage, drop them on the family dog or cat passing below. Dicker says that another favorite joke is to defecate on the other pets (which lends credence to Gary Larson's *Far Side* cartoon in which a bird flying above a car that has just been spotlessly cleaned looks down and says, "You're mine, you're all mine").

As these stories suggest, the line between humor and just plain mischief is blurred. With chimps, the line is often nonexistent. Some years back at the Tulsa Zoo, the keepers at the Chimp Connection exhibit went off to a Thanksgiving party secure in the knowledge that painters working on one part of the indoor cage area were separated from the chimps by bars and plywood. The party was interrupted when they got an urgent call to get back to the exhibit. Amy Morris returned to find that the chimps had somehow appropriated the plywood and the protective tarps, as well as paint, brushes, and gloves from the terrified painters. Facing her was a tableau of mothers and babies, the babies completely covered with white paint, and one of the mothers wearing one of the painter's gloves. The plywood was nowhere to be seen.

Furious, Amy said, "I want that plywood now," causing

the panicked chimps to scramble to look for it. They could not figure out how to get the plywood back through the door, but they brought every imaginable object as peace offerings: the tarp, sticks, and anything they could find on the floor of the cage. Moreover, chastened by Amy's sternness, they cleaned up the paint-covered babies. When the chimps came back later, not one baby had a speck of paint on it.

Finally, one word or two about training and the games animals are taught to play by humans. With sensitivity and insight, animals can be taught to do astonishing things. Some of these routines, like the arithmetic performed by Clever Hans, may have simple explanations, but sometimes lost in our analysis and appreciation of these tricks is what they mean for the animal, and indeed whether the animal enjoys doing them. Sally Blanchard has taught one of her parrots, a caique named Spike, a routine in which she pretends to shoot him with a finger pointed like a gun. When she says "bang," Spike falls over on his back and sticks his feet in the air, looking for all the world like the cartoon version of a dead bird. When they do this routine at parrot shows (Blanchard speaks frequently on parrot behavior around the country), people applaud, and Spike will respond by spreading his wings, puffing out his chest, and strutting around the stage. He also does flips as part of his comic routine and hops around on two legs at once as though he were on a pogo stick.

There is little question that Spike enjoys performing. Sally says that the more applause Spike receives, the more he swaggers around. Apart from producing a reaction in people, a routine has other satisfactions. Learned routines offer relief from the monotony of captivity, of course, and they also in some cases help seal bonds with a trainer or keeper. In this sense, a routine can be like an elaborate greeting ritual in which human and animal renew their relationship.

Something like this may have been at work in a remarkable incident caught on film in which a retired trainer and his performing elephant were reunited after a fifteen-year hiatus. The trainer, Charlie Franks, had bought the female as a baby in 1955. He named the elephant Nita, and she performed in a traveling circus in Southern California for fifteen years before retiring in the early 1970s. In 1989 Charlie came to see Nita at the urging of a reporter, Huell Howser, from station KCET in Los Angeles. Standing near the tall fence that formed the boundary of the elephant enclosure, Charlie called Nita, saying, "You'd better get over here." Hearing his voice, Nita put her trunk in the air and quickly came over to the fence. She ran her trunk all over Charlie, then opened her mouth for some jelly beans, the old reward Nita had gotten when they performed. Next Charlie got her to run through parts of her routine like sitting up and doing a headstand. Afterwards the reporter asked Charlie if it was easy to get attached to an elephant. He answered with another question: "Is it easy to get attached to a baby?"

There is no question that these routines become an important part of an animal's life. Gail Laule remembers being called to consult on the case of a former circus elephant named C'sar now housed at the Zoological Garden in Ashboro, North Carolina. The bull elephant had been taken out of performances because he had become too aggressive. At the point Laule met him, no one had worked with him in three years. Using reinforcement techniques that gently encourage an animal to touch a target and then reward it for results, Gail calmed the elephant and also got C'sar to go through his whole routine. The elephant responded so enthusiastically that he then did five more tricks than the trainers had asked for, and kept trumpeting for them to come back for more.

At the bankrupt Marineland south of St. Augustine, Florida, the trainers continued to put dolphins through their routine three times a day, even though the facility had been closed for four months. As reported by Jim Carrier in the *New York Times*, the keepers continued leading the dolphins through complicated aerial twists and flips because they liked it, and because without the release of these play sessions, the trainers feared that the park's nineteen dolphins might become aggressive towards each other.

According to the report in the *Times*, three of the dolphins performed a routine that they had come up with themselves. This too is not surprising considering Louis Herman's "tandem creative," mentioned earlier in the book. In the 1960s, Karen Pryor, Richard Haag, and Joseph O'Reilly published a paper entitled "The Creative Porpoise: Training for Novel Behavior," in which they showed how through the use of operant conditioning a trainer could teach dolphins to introduce new elements in a routine. It is worth looking deeper into the example cited in the introduction, which took place three decades later, to explore how the dolphins figure out what routine they will do together, and what it might mean.

The University of Hawaii's two dolphins circled underwater until they were swimming together, then leapt out of the water side by side, simultaneously spitting jets of water before resubmerging. Dolphins do not ordinarily swim with water in their mouths, and so, somehow, had agreed to take water in their mouths *before* jumping into the air. But how?

Lou Herman's study is concerned with the degree to which the dolphins can understand the syntax and words of the invented language he has taught them, and not what they say to each other, and so he is unwilling to speculate on how they communicate with each other when deciding what trick to perform. It is possible that dolphins have their own language and grammar and simply discuss what they are going to do, but as of now, there is no evidence that they have developed such abilities. It is also possible that they have some nonsymbolic way of conveying their intentions and ideas to each other. In the absence of evidence, it is hard to say how they communicate underwater.

According to Karen Pryor, the reductionist explanation is that the tandem creative is an example of exquisitely timed imitation. Dolphins are highly mimetic, and it could be that while underwater one animal opened its mouth to take in water while looking at the other in a "Do what I do" gesture. These are sophisticated animals that have done routines for humans thousands of times, and, she argues, it would not take representational language for one animal to convey to the other, "Boy, have I got a great trick for them today, let's do this. . . ." Maybe Pryor is right, or maybe the dolphins do have some symbolic facility that lets them plan tricks together.

Pryor does not doubt that these and many other animals enjoy these routines. "These animals evolved to act in the environment and make it work for them through their actions, and it is reinforcing to get a reward as a result of something you do," she says, using the language of behaviorism. Switching to the language of welfare reform, she continues, "It's much more reinforcing for a cat to hear a rustle in the grass and then pounce and catch a mouse than it is to have its mom drop a dead mouse in front of it." In short, there are emotional rewards for doing something well and getting paid for it.

While the principles of operant conditioning may seem dreary and mechanistic, the enthusiasm of the animals being trained is palpable and real. Max and Patty, the two elephants that play hide-and-seek at the Bronx Zoo, get excited when they are getting "warmer" looking for some hidden object. Lou Herman's dolphins, Akeakemai and Phoenix, get excited and race back to the trainer when they get something right. When wrong, they sometimes take out their frustration on the object they are asked to identify or move, whacking a hoop or basket as though it were the object's fault.

But to focus on tangible rewards is to miss the point. These animals know they will be fed. For many animals in captivity, the real reward is the game itself—to which I would add that the sentience and sense of fun on the human side of these games is reciprocated to some degree by the animal involved.

"SHE DIDN'T KNOW HUMAN, AND HE DIDN'T KNOW GORILLA ..."

Trade and Barter

L EST there was ever any doubt, the 1980s and 1990s have underscored the importance of money, trade, and barter in modern material culture. Indeed, in recent years concerns of Mammon have eclipsed other aspects of human nature to the point where it is easy to believe that money is the defining attribute of human culture, and easy to forget that *Homo sapiens* existed for many tens of thousands of years before the need to store foods gave rise to banking and markets. While it is unlikely that scientists will ever stumble upon a group of chimps trading banana futures, in captivity various animals turn out to be enthusiastic traders. Some captive animals seem to have a grasp of money and savings, and a few have even risen (or fallen) to the level of understanding the basis of inflation and counterfeiting, not to mention extortion and theft.

In the wild, some species demonstrate various precursors of the consumer society. Squirrels and a number of other animals store food. In chimp colonies, males have been observed trading food for sex, and there are even examples of conspicuous consumption and thievery. Hyenas, for instance, will study the habits of humans in order to pick the best time to steal food, but they also steal things of no known utility to hyena life. Scientists have found pressure cookers and binoculars in hyena dens.

Much of the protoeconomic activity that has been observed in the wild does not require awareness. The squirrel who stores nuts for the winter does not have to know about discounted rates of return or opportunity costs. Nature works that out for the squirrel over eons and encodes the results in the animal's genes. While much of the sharing, gift-giving, and trading that goes on in the wild can be explained without invoking awareness, in captivity there is a mountain of evidence that the more sophisticated animals quickly learn concepts fundamental to any economic system.

Animals to some degree must become students of the humans who control their lives. The great apes in particular are exquisitely sensitive to changes in the power balance of any given social situation, especially when circumstances change in a way that temporarily might give them the upper hand. One such occasion is when something rolls into their exhibit or is left behind. The more worldly animals recognize the concept of value, particularly when it is defined as "something I have that you want." Once animals have what a human wants, the balance of power shifts, and gorillas, orangutans, and a number of other animals are not above exploiting such occasions for all they are worth.

Charlene Jendry of the Columbus Zoo says that captive apes develop a trader's ability to see what the market will bear when they get their hands on something a human does not want them to have. She recalls one day when a volunteer came looking for her, saying, "Colo took something from one of the babies and has it in her hand." Charlene came out to the exhibit, but she could not see what Colo had. Charlene offered Colo some peanuts, only to be met with a blank stare. Charlene, realizing that they were negotiating, upped the ante and offered a piece of pineapple. At this point, Colo took the gamble of tipping her hand, and without making eye contact with Charlene, she opened her hand and revealed that she was holding a key chain, much in the manner that a fence might furtively show a potential customer stolen goods on the street.

Relieved that it was not something particularly dangerous or valuable, Charlene gave Colo the piece of pineapple. Astute bargainer that she was, Colo then broke the key chain and gave Charlene a link, clearly figuring, "Why give her the whole thing if I can get a bit of pineapple for each piece?" Because she really did not care, Charlene did not tackle the daunting task of convincing the gorilla that the key chain was more valuable intact.

In fact, while Colo has learned well how to increase her leverage by giving back treasures in parts, she has not quite figured out all the relative values that humans assign to objects whole or in parts. On another occasion when she got hold of a plastic cup that Charlene wanted back, she tore the cup in half and again bartered for food, piece by piece.

Kyle Burks, formerly of ZooAtlanta and now of Disney's Animal Kingdom, says that ZooAtlanta had several trading gorillas, but none more astute than Ivan. This gorilla received his food through the mesh. He would hold out his hands, take the food, and then, like many children, proceed to eat everything he was given except for the green beans. These he would collect in his hand, and then try to trade them back to the keepers for grapes.

Ivan did not meet another gorilla until he was close to thirty years old. He had been raised in Tacoma, Washington, and lived with a human family until he was five or six and began wrecking their house. Before he left, however, they had taught him to eat with a knife and fork, and also to clean up after himself. "Give him a towel and he would begin wiping up," recalls Burks. "We could almost get him to clean the cage by pointing to banana peels and other garbage and asking him to bring it to us." Ivan is still so fastidious that he makes slippers out of clumps of hay so that his feet won't get wet when the floor is damp after cleaning. In any event, Ivan learned to trade in his life before meeting other gorillas, a circumstance that led to a number of unfortunate encounters when he was getting to know other members of his species.

"He didn't know that the rules of trading he had picked up from humans did not apply to gorillas," said Burks. Ivan went ballistic when he handed a Frisbee to a female gorilla, probably expecting a treat in return, and instead she ate it. This was just one of a number of crossed signals he had with his own species. Once he stuck his tongue through the mesh, trying to kiss the female, and she punched it, causing Ivan to "go berserk," in Burks's words. She held out a hand in typical gorilla reconciliation behavior, but poor Ivan, not having encountered this particular gesture, punched her hand, causing the female to throw a fit. "The problem," says Burks, "was that she didn't know human, and he didn't know gorilla."

Orangutans are no slouches at trading themselves. Helen Shewman at Seattle's Woodland Park Zoo says that they even know the word *trade*. Towan, the dominant male, invented his own version of trading piecemeal. He got hold of a piece of hose at one point, and was slowly letting Helen pull it out, bit by bit, in return for peanuts. When he got to the end, he seemed to realize that the gravy train was about to end, and so he took the simple expedient of pulling the entire hose back into the cage so as to begin the negotiations anew.

Sometimes the desire to gloat overwhelms the orangutan's natural trading instincts. Helen had left a Tupperware container full of Cheerios in Towan's cage. He took it, came up to the bars, and with a flourish produced the container from behind his back with one hand. "It was," says Helen, "as though he was saying, 'Oops, look what you forgot!' " To drive home his point, he opened the lid, poured two quarts of Cheerios into his mouth, and then casually tossed the container over his shoulder. It's possible, of course, that Towan had made a cost-benefit analysis and decided that the joy of thumbing his nose at Helen and the treat of two quarts of Cheerios was more valuable than anything Helen might offer in return.

It must also be said that Towan was capable of great generosity as well. Sometimes when he saw Violet Sunde smoking (it's amazing how many zookeepers smoke), he would get a piece of paper, put hay in it, and try to tie it together with a piece of grass (orangutans are gifted knot-tiers). Then he would hold it in his lip, and when Violet approached, he would come somersaulting over and give it to her.

Towan also proved capable of great tenderness. When a baby opossum crawled into his cage, Towan picked it up by the tail. Libby Lawson, the keeper on duty, feared the worst— a gruesome dismemberment in front of visiting schoolchildren as Towan traded pieces of the opossum for treats—and so she got some special treats to plead for the return of the opossum whole. Her fears proved unfounded. Despite the fact that the baby bit him, Towan never let go, and never let the other curious orangutans touch the baby. Instead he brought it over and gave it unharmed to a vastly relieved Libby.

Unlike Colo, Towan seemed to respect the integrity of inanimate objects that humans consider valuable. Helen once discovered that she had left her radio handset in the exhibit when she saw Towan inspecting it. He carefully removed the antenna and then screwed it back in. Then he held it to his ear and tried to use it as Helen did, although he could not figure out how to turn it on. Helen, laughing, envisioned him delivering his demands: "I've got Helen hostage. I want two hundred pounds of bananas and a boat to Indonesia."

Mike Yznaga, an orangutan keeper who formerly worked at the Topeka Zoo, says that the orangutans there were inveterate traders as well. They too figured out that if a keeper dropped three keys, it was better to barter them one by one than agree on a price for the whole lot. Unfortunately for the more junior-ranking apes, the dominant male and female seemed to have learned their business methods from the Mafia. They nominated themselves as the chief traders, and no matter which orangutan found some item for barter, either Jonathan or Patty

would seize it through force or threats and then take it to the
keepers to see what the market would bear that day.

Geoff Creswell, who worked with Yznaga, recalls that if
keepers did not accidentally drop any keys, the orangs
would resort to the expedient of stealing them in order to
have something to trade. "Unlike chimps, who will try to
steal your keys every time you walk past," says Creswell,
"an orangutan will make the attempt only if he or she is certain that he or she will succeed." For the most part, the apes
would then trade the keys back for grapes. On rare occasions, they would refuse, and if the keys were needed, Geoff
or one of the other keepers would threaten to douse the
orangutans with a hose. He remembers fondly one true rebel
who walked right through a stream of water and shook the
bars separating Geoff from her. Then she decided that
enough was enough and threw the keys at Creswell. Thereafter, all Geoff had to do was say, "I'm going to get the hose,"
and she would return whatever it was that she had stolen.

Even in the absence of objects to trade, some animals have
figured out how to extract bribes by foiling the wishes of their
keepers. Before the gorillas at the Bronx Zoo in New York
moved to their new home in the Congo Gorilla Forest, the
routine was for one keeper to lure them inside at the end of an
outdoor session, then yell out "Shut the door" to cue another
keeper outside to close the gate between the indoor and outdoor areas. Denise Smith, a senior wild-animal keeper at the
zoo, says that Huerfanita, an adult female, figured out what
"Shut the door" meant, and would run down and block the
gate from closing whenever she heard the words. "We would
have this ridiculous situation with a bunch of humans trying
to bribe this gorilla with bananas to come inside." Denise's
solution has been to resort to code words like "The fat lady
sings," or to say "Please shut the door" in a sweet conversa-
tional way in order to fool the gorilla. (This suggests that
Huerfanita was reacting to tone of voice and context rather
than decoding the actual words the keepers were speaking.)

There is some evidence that apes understand the utility of giving bribes as well as taking them. Hugh Bailey, now at the Woodland Park Zoo, previously worked with gorillas at the Dallas Zoo. While he was there, the zoo was in the process of moving Fubo and Demba, a male and female gorilla couple, to a new site. Once there, however, Demba refused to go outside. After a few days of this, Fubo came in from the outside, brought in an orange, set it down, and then went out again. Says Hugh, "It was clear to us that he purposefully left it for her." The message seemed to be, "What does it take to get you outside?"

There is also evidence that chimps, orangutans, and even some monkeys are capable of moving beyond trading to the more abstract domain of dealing with money. Hal Markowitz is one of the pioneers in attempts to "enrich" (which in essence means to make more stimulating and interesting) the lives of captive animals. Working at the London Zoo in the 1970s, Markowitz set up a plastic-token dispenser in the enclosure that held Diana monkeys. The monkeys could take tokens out of the dispenser to use in a machine that dispensed food. The Diana monkeys demonstrated both bankerly and endearing traits in the ways they used these tokens. Some would store up tokens, which shows a willingness to postpone gratification—the very heart of the Puritan ethic, which many humans have a hard time matching. The monkeys would not use tokens to get food for each other, but one would handle the tokens for its mother, who, like many old-fashioned mothers, had never figured out how to use the newfangled machines.

Almost all animals will perform for a reward, but some come close to understanding the concept of employment. David John Shepherdson, a consultant on enrichment programs, notes that Charlie, a male chimp at the Portland Zoo, will go out and round up the female chimps and bring them in from their outdoor enclosure in return for a reward. The next step, of course, is to determine whether an animal will work for money which it later can spend on treats.

It turns out that at least apes enthusiastically embrace the concept of working for wages. Chantek learned about money and wages as well as sign language during his years at the University of Tennessee. In the course of her long-term study, Lyn Miles filmed Chantek as he used his vocabulary of 150 signs in various ways. One episode shows Chantek making the signs "go ride" as a way of requesting a ride in the car. Lyn responds by telling him in sign language that first he needs money. This prompts Chantek to go get the "money" he had earned earlier by cleaning up his room. Once he has paid for the ride, the two head for Lyn's car.

Chantek's understanding of the money economy seems to have extended far beyond simple transactions to such sophisticated if morally dubious concepts as inflation and counterfeiting. Lyn first used poker chips as the coin of the realm, but Chantek figured out that he could extend the money supply by breaking the chips in two. When Lyn switched to using washers, Chantek would find pieces of aluminum foil and try to make imitation washers that he could pass off as the real thing.

He also learned more virtuous uses of money, such as saving and charity. Lyn says that one of her classmates when she attended Yale was a Rockefeller, a family that knows a thing or two about how to deal with money. Miles remembers her classmate remarking once on the guidance she received. She told Lyn that at family meetings with her cousins when she was a child, she was encouraged to save 10 percent of her allowance, give 10 percent to charity, and spend the rest. Lyn decided that what worked for the Rockefellers might work for an orangutan, and she started encouraging Chantek to save.

I got to see Chantek use his money system when I visited him at ZooAtlanta, where he now lives. Lyn brought to his cage a bucket of fruits and vegetables and a cup full of washers. After some greetings, Lyn asked Chantek, "Where money?" He responded in sign language, "In cup." Chantek then asked for money, and Lyn said, "First you have to do work." She then asked him to tie a knot. Chantek looked around his cage for a string but came up empty. So Lyn offered him some string that keeper Laura Mayo had produced. Using his lips and fingers, Chantek slipped the string around the wire mesh of his cage and secured it with a half hitch. (Lyn says that if given beads, Chantek will make a series of knots to space the beads and present her with a necklace.) Chantek got two coins for this, one of which he spent on a tomato. The earning, trading, and spending went on until Chantek found himself down to his last coin. He asked for another piece of tomato, and Lyn asked for money in payment. Faced with a cash-flow problem, he visibly hesitated, loath to give up the last of his savings. In the manner of a free-market Republican, Lyn responded by encouraging him to earn more money.

They then played a game of Simon Says in which Lyn made a gesture and then paid Chantek if he produced a good approximation of it. After Lyn poked a piece of straw through a piece of paper, Chantek took the piece of paper and, using his mouth, poked the stiff grass through the paper (in so doing, he provided a nifty example of program-level imitation, a sophisticated form of imitation, in which a person or animal understands how a series of movements serves to achieve some outcome). Then Lyn decided to try something new. She said to Chantek, "Simon says do what Eugene does." Taking the cue, I stood up and made a big circle with both arms. Chantek, leaning on one elbow and looking at me with a mixture of curiosity and disdain, made a dismissive and halfhearted half-circle with his free arm. Lyn told him he could do better than that. Prompted by his need for funds, Chantek heaved himself to his feet and then made a gigantic circle with his enormous arms.

I did not see any evidence that Chantek was giving 10 percent of his earnings to charity, although I did see the orang-utan offer an example of sharing that a robber baron might envy. When Lyn gave Chantek some grapes and asked him to share them with her, Chantek promptly ate all the grapes.

Then, seemingly remembering that he'd been asked to share, handed Lyn the now bare stem.

Animals also seem to understand that rewards work in both directions. Some years ago the dolphin specialist Karen Pryor, who became a zoo consultant on operant conditioning, gave the staff of the National Zoo instruction on how to use food rewards in training. As reported in her book, *On Behavior*, one of the zookeepers who adopted her techniques was Melanie Bond, who now works with Rob Shumaker at Think Tank. First Bond taught one of the young orangutans named Junior to clean up his cage in return for the opportunity to play with her whistle (an object of orangutan fascination). Then she began using food rewards to try to bring a chimp named Ham out of his lethargy. Ham had achieved renown in the 1950s as a pioneer of space travel. His successful flight 157 miles into space aboard a Mercury-Redstone rocket, and his return alive, bolstered confidence that humans could survive the rigors of flight. Melanie got Ham out of his shell by rewarding him for various activities, and very quickly he grasped the essentials of the transactions. He also seemed to grasp the element of reciprocity as well as the concept of reinforcement. According to Pryor, one time when he saw that Melanie was going to open the door to his outside cage, he rewarded her with a piece of celery. Maybe he was just imitating what she typically did in these circumstances— or perhaps he thought that with enough celery, he might encourage her to open the door more often.

Kanzi, a bonobos who has been part of a long-term language experiment conducted by Sue Savage-Rumbaugh at the Georgia State University Language Research Center, also could earn spare change by doing well in practice sessions. On one occasion, I watched Rose Sevcik drill the pygmy chimp on his vocabulary. When he correctly identified eggs, Rose gave him "money" in the form of a token. He then used his money to buy an orange. Correctly identifying clover earned him another token. This time he bought some raisins.

Since apes, like real estate developers, seem to be able to grasp the concept that they can make more money by subdividing something rather than selling it whole, it is worth pondering how much math different animals can handle. Tetsuro Matsuzawa of the Primate Research Institute of Kyoto University taught a female chimpanzee named Ai to discriminate between Arabic numbers. The idea was to see whether the chimp would use, without prompting, the numbers in an internally consistent way, thereby providing evidence of some nascent syntactic abilities.

When I met Ai in the early 1990s, she had proven capable of determining whether there were eight or seven dots on a screen, or six or three, and so on. The way this worked was that Matsuzawa would show Ai two random arrays of dots on a video screen, and she would have to choose the appropriate number for a cluster. When I met Ai, Matsuzawa was trying to teach her the number ten, which would have represented a record for the animal kingdom, at least in formal studies.

There have been numerous experiments with other species as well. In the 1940s and 1950s, various experimenters taught crows to distinguish between numbers up to eight. Irene Pepperberg has taught the parrot Alex to correctly identify numbers as high as six. She will show Alex a random assortment of objects—there might be six red pieces of wood (Alex says "rose" for red because he has trouble with the letter *d*), four green puff balls, and five pieces of green wood—and then ask him, "How many rose wood?" A good deal of the time Alex will get it right, answering in his sweet child's voice, "Six." (Pepperberg says that Alex has an Eastern accent with some Midwestern inflections.) When I caught up with Irene in Denver in 1999, Alex was still stuck on six.

Alex has learned numbers by looking at collections of objects. His ability to count accurately the number of one type of object in a clutter of objects that are similar in some ways but different in others is impressive. It would be more impressive still if Alex could say that six is larger than five without any objects present, or that if you put three green woods and two rose woods together, you would have more in wood numbers than the four puffballs. But that has not happened yet.

It should not come as a surprise that even the brightest animals have trouble with numbers beyond single digits. Mathematics has been described as the language of thought, and computing sums in one's head places a fantastic burden on memory and concentration. Ai was fast as lightning in picking the appropriate answer, but this does not necessarily mean that Tetsuro Matsuzawa was unlocking some latent mathematical ability, or tapping into mathematical abilities already in use in the wild. Marc Hauser, who studies animal consciousness at Harvard University, points out that it was just as difficult for Ai to learn to discriminate between four and five as it had been to differentiate two and three. Human children tend to "get" the concept of numbers after figuring out the first few, and their learning curve rises much more steeply. Ai is learning numbers, to be sure, but Hauser says that the arduous way she has acquired them suggests that her internal representation of those numbers is likely to be different from that in humans.

We have been equipped by nature for tasks like juggling numbers, but these signal human abilities may also be present in more limited form in our closest relatives, and in other higher mammals. Frans de Waal suggests in his book *Good Natured: The Origins of Right and Wrong in Humans and Other Animals* that trading, sharing, and gift-giving in captivity are more likely to occur in animals that engage in these activities in the wild. Other aspects of the money economy also seem to have wild antecedents.

This makes sense, but it is not the whole story. Orangutans, who in the wild exchange material favors with nowhere near the frequency of chimps, turn out to be gifted *hondlers* in captivity. Don't expect to see an orangutan trading derivatives on the floor of the futures market in Chicago anytime soon, but they more than hold their own in the primitive bazaar of daily life in the zoo.

AH, TREACHERY

Deception

The Prince who wishes to become King
should be sure to possess all the good
qualities and be compassionate, but he
should also be ready when circumstances
require that he act in his own self-interest.
—Machiavelli, *The Prince*

WHEN news of some act of human barbarity, greed, venality, or deception prods us towards misanthropy, we often turn to animals, projecting onto them a moral superiority to humans, the fallen beings who have tasted the forbidden fruit. *Innocence* is the word most often applied to animals, implying a guileless existence in which they always reveal their true feelings. How many times have we heard an animal trainer or pet guru say, "An animal will never try to deceive you."

I don't think so.

Clearly these trainers and other sentimentalists have never encountered the orangutans and gorillas at the Woodland Park Zoo, who lie about whether they have received their afternoon treats, or Alex, the parrot who tricked Irene Pepperberg into getting close enough to be bitten, or my cat Junior, who is always pretending that she hasn't been fed when she has. To their credit, animals don't deserve their reputation as moral paragons. I say to their credit because the more sophisticated forms of deceit require consciousness, an awareness of others' mental states, and the propositional abilities that go with artful scheming. Conversely, if some animals possess higher mental abilities, they should also be capable of putting those higher abilities to lower ends. Far from an innocent kingdom populated by straight shooters, the animal world is rife with con artists, devious manipulators, and dissemblers. The spectrum of deceptions ranges from misleading signals produced by the blind forces of evolution to con jobs that require an animal to be capable of something akin to the mental chess played by spies as they try to plant false beliefs in their opponents' minds.

Rob Shumaker remembers one female gorilla at the National Zoo who developed strong feelings about various keepers. She was loyal to those she liked, but very aggressive towards those she did not. On one occasion, she casually walked up to the wire mesh that separated her from one of the keepers she did not like. Using a come-hither gesture, she beckoned the man to come closer. When the keeper approached within range, she suddenly pulled out a stick she had hidden behind her back and tried to stab him.

At the Atlanta zoo, gorillas have adopted similar tactics. They will put an arm through the bars, reaching after someone, and appear to be straining, as though they can only reach so far. When the unsuspecting subject comes closer, their reach suddenly and magically extends and they grab their victim's shirt.

The ways in which apes try to deceive humans are analogous in some ways to the tactics they employ on each other in the wild. Chimps and gorillas prefer to accomplish their goals through bluster and mock charges rather than real fights. "Deception is a huge part of the gorillas' behavioral repertoire," says Disney's Animal Kingdom's Kyle Burks. "Look at the degree to which bluff is part of the daily life."

Even parrots will use come-hither tactics to get an unsuspecting human within range for an attack. This happened to Irene Pepperberg after she had been away from Alex for a number of weeks. Shortly after she greeted him, he wanted a tickle, something, she says, he never asks for himself. Thinking that he needed reassurance after her long absence, Irene walked over and began ruffling his feathers. No sooner had she gotten within range, however, than he wheeled his head around and sliced open her finger with his beak. Apparently, he still harbored some resentment over her long absence. Then the one-pound bird said, "Sorry."

A somewhat larger animal, the elephant, will also resort to treachery to avenge some perceived slight. At the Tulsa Zoo, the keepers try to ensure that there are no loose rocks lying around, a lesson they have learned through painful experience. Sneezy, a twenty-seven-year-old bull elephant, has a habit of concealing stray rocks in his trunk when a keeper is not looking. Then, says Mark Swanson, he will hurl the rocks with his trunk, trying to plaster the keeper.

Conversely, a number of species will disingenuously use body language to get more attention from humans, or, perhaps, simply to see how they will react. Heapo, one of the male dolphins in Louis Herman's language experiment, used to wrap himself in rope and pretend that he was in trouble. The rope would fall off, showing that it was a ruse, but Heapo would still act distressed. Heapo most likely had to realize that the rope had fallen off, so perhaps this was his way of showing the trainers that they had been the victims of a practical joke.

Helen Shewman, of Seattle's Woodland Park Zoo, recalls another form of deception adopted by the orangutan colony to get more food. Every day, using a porthole that provided access to the exhibit, keepers would give the apes snacks in the form of oranges dropped through the opening. Helen remembers that one day she dropped an orange for one of the females, Melati. Instead of moving off, Melati looked Helen in the eye and held out her hand. Thinking that the orange must have rolled off somewhere inaccessible, Helen gave her another one. Only when Melati moved off did Helen notice that she had hidden the original orange in her other hand.

Towan, the big male, had watched this whole charade, and the next day he too looked Helen in the eye and pretended that he had not yet received an orange. "Are you sure you don't have one?" Helen asked. He continued to hold her gaze and held out his hand. Relenting, she gave him another, then noticed that he had been hiding his orange under his foot. Even as he walked away he tried to keep up the deception by sliding his foot along the floor, attempting to conceal his contraband. Usually, says Helen, Towan simply takes the orange and walks away.

At the same zoo, Nina, one of the female gorillas, used a similar trick to get an extra apple. The keepers would give the gorillas an assortment of fruits and vegetables each afternoon, and on this particular occasion Judy Sievert tossed Nina an apple, which rolled away. Instead of going to get it, Nina just "sat there sadly," in Judy's words. Judy continued her rounds, handing out yams and apples to the other gorillas, but Nina sat there looking appleless and downtrodden. Taking pity, Judy tossed her another apple. As soon as Nina had it, she got up and went over to where the first apple had rolled away, taking it too.

Chantek, the orangutan who was part of the long-term study of language undertaken by Lyn Miles at the University of Tennessee, also used various forms of deception. One trick was to point at something, and then, when an unwitting visitor's attention was diverted, try to take things from the visitor's pocket. Lyn says that he would also steal cookies from the kitchen, then try to rearrange the remaining cookies so that it looked like nothing had been taken.

Charles Horton, a senior gorilla keeper at ZooAtlanta, says that Willy B., a wily forty-one-year-old male gorilla, used to enlist him in what amounted to a con to get treats from zoo visitors. If Willy B. saw that a visitor was chewing on a stick of gum and Charles was nearby, Willy would come up to the edge of the enclosure and pretend to be chewing. The visitor would say, "What's he eating?" Playing his role in the shakedown, Horton would say, "He's not eating; he's telling you he wants a piece of gum." Responding to the prompt, most visitors would then say, "I've got another piece," and as soon as they reached for their pockets, Willy B. would head off to the place where food was delivered to wait for Horton to arrive with the gum.

Brad Andrews, the director of zoological operations for Busch Gardens in Florida, remembers one artful piece of dolphin sneakiness from when his days at Marineland. A large pool housed a community of dolphins, and the marine mammals loved to play with objects carelessly dropped by humans, things like ballpoint pens, quarters, and pocket protectors. Periodically, the management would send in divers to clean up the debris. They would do this by methodically sweeping the bottom with a vacuum device. On one such sweep the divers were finding very little. The reason was clear to anyone observing from the top. One of the older dolphins named Zippy was going ahead of the divers, just beyond their range of sight, picking up objects in his beak. Then he would swim around behind the divers to an area he had already swept and which they could not see. There he would redeposit the treasures.

Once the cleaners got wise to Zippy's game, he tried a new ruse. When he saw teams of divers getting ready to enter the pool and clean, he would quickly pick up his favored objects and hide them beside the water jets in a spot where the water flow would not carry them away but where the cleaners could not get to them.

One of the most hilarious examples of deception is also one of the most ambiguous. Of all the scientists exploring language, Penny Patterson has been the most aggressive in asserting that the gorilla Koko has used her sign vocabulary to convey complex ideas. Penny's argument that one of her gorillas used sign language to talk about being captured as an infant in Africa inspired Michael Crichton's book *Congo*, but this, and Penny's assertions that Koko uses the language to rhyme and talk about death, have been met with extreme skepticism by critics who accuse her of wildly overinterpreting the gestures Koko makes.

There is no question, however, that sign language is very much a part of Koko's life. She signs all the time to make her wishes known, and not just to Penny but to anyone who comes by. She may also sign to escape blame. The incident in question took place over twenty years ago when Koko lived in a trailer on the campus of Stanford University (she has since moved to Woodside, California). Then a rambunctious juvenile, she inadvertently knocked a sink in the bathroom off its moorings. Penny had a number of assistants who would tend to Koko and teach her signs, and at the time a slight young woman named Kate was in the trailer. When Penny saw the damaged sink and confronted the gorilla, Koko pointed to the sink and signed, "Kate there bad." Was Koko throwing out a wildly implausible accusation? (After all, if there is a broken sink and a gorilla in the room, who is going to believe that the perp was a small young woman?) Or was she randomly ticking off the items relating to the incident, with no intent to deceive?

No less than gorillas, orangutans also seem to realize that language offers great opportunities to shift blame and otherwise deceive those around them. A young male orang can be obstreperous, and on numerous occasions Lyn Miles would confront Chantek over some piece of misbehavior. In a typical encounter she would say, "Bad Chantek, what you do?" Chantek's standard reply was "Good, good Chantek," to which Lyn would then respond, "Then who did?" Chantek would then blame various human assistants, and on one occasion he unchivalrously tried to shift blame to "Kitty," a stray cat he had adopted.

In nature, deception runs the gamut from trickery that is genetically encoded to sophisticated ruses knowingly perpetrated. In the former category, for instance, the Rafflesia of Borneo, which at three feet across is the world's largest flower, evolved to look and smell like rotting meat in order to trick carrion-eating flies into serving as its pollinators. Another born liar is one particular moth caterpillar of Meso-america, which has a colored pattern on its underbody that makes the bug look for all the world like a viper. Over time, some animals have learned that it is in their interest to give false signals. For instance, the white-winged shrike tanager serves as a sentinel for other birds in the western Amazon. It provides an early-warning system through its vigilence, but according to ornithologist Charles Munn, the bird also takes advantage of its position by occasionally making the alarm call when no hawk is around. As its feathered colleagues head for cover, it wolfs down all the food in sight.

Another bird that resorts to deception is the zone-tailed hawk, which will imitate the flight pattern of a vulture, com-plete with rocking wings, in order to lull macaws or para-keets into a false sense of security. (Birds have nothing to fear from a scavenger.) Once the camouflaged predator has gotten overhead, however, it swoops down to attack the smaller birds.

Some birds will try to deceive other species into raising their young, shifting the burden of food gathering for vora-cious nestlings to unwitting dupes. One of the ultimate free-loaders might be the cuckoo. A female cuckoo will lay her egg in another bird's nest alongside its eggs. When the eggs have hatched, the young cuckoo will kick the other chicks out of the nest so that the unsuspecting mom can devote all her energy to feeding and raising the little interloper.

None of the examples I have cited so far require aware-ness on the part of the deceiver, but they do raise a question. If dishonesty can get you a good meal or someone else to raise your kids, why be honest at all? Why isn't the animal world an anarchic sink entirely populated by grifters and Fagins?

In fact, all species have evolved ways of keeping decep-tion and graft within reasonable bounds (we humans are still working on the problem). Marion East and Heribert Hofer, two German wildlife biologists, have looked into this question with regard to the hyena, a highly social animal of the African plains that has come, undeservedly, to symbolize the human attributes of treachery and viciousness.

Hyenas inspire fear, awe, and hatred in people, and perhaps for this reason the predators figure prominently in the myths and rituals of Africa. In Tanzania, shamans sometimes ride on the backs of tamed hyenas to meetings. In Ethiopia, the men in Harare reportedly have a ritual in which they allow wild hyenas to take meat out of their mouths.

Despite their reputation, hyenas can be remarkably civilized. Purportedly, one pet hyena owned by a wealthy family in France developed a fondness for chamber music. When it heard the sound of live string music, it would scratch at the door and come in and lie down peaceably in front of the musicians, unnerving those not accustomed to having an immensely powerful carnivore in the front row. "These are social animals, in behavior somewhat like wolves, and they readily understand that you have certain rules," said Hofer.

In the wild, the animals have evolved rules governing how much cheating a group of hyenas will tolerate. The predators live in extremely large groups of up to fifty to sixty animals. They hunt in groups, and it might take a half dozen to stalk and kill a zebra. Even after expending all this energy on a kill, the hyenas might lose their spoils up to 20 percent of the time as lions move in to steal the carcass. Understandably, then, hyenas have a specific alarm call that alerts the group when a lion is approaching. Once the alarm is sounded, a common hyena response is to head for cover. The opportunity this presents does not pass unnoticed among more unscrupulous hyenas. They will sound the alarm call and, when the hardworking hunters scatter, move in to steal a few quick bites of meat.

This kind of thing could easily get out of hand, but in the patient course of evolution, hyenas have worked out how much cheating can be tolerated without jeopardizing the viability of the group. According to Hofer and East, hyenas will tolerate a certain number of false alarms. Hyenas also will heavily discount alarms sounded by nervous females who give the call all the time, suggesting that they have a scale of credibility for different members of the group. The lesson, as one behavioral ecologist put it, is that an animal must be frugal with deceit. The white-tailed shrike tanager only gives the fake alarm call 10 to 15 percent of the time, presumably a tolerable amount of graft worked out over time by nature as it balanced the risk of ignoring the alarm call against depredations on the common food supply.

Beyond this, nature tends to keep cheating and dishonesty within limits by making it costly in one way or another. Everyone is familiar with the archetype of the charming rascal who presents himself to eligible young females as more solvent than he really is. It turns out this is a problem that extends across species, and some have evolved elaborate defenses against such deception.

Gail Patricelli, a behavioral ecologist at the University of Maryland, has been studying a spectacular example of gift giving in the mating ritual of the satin bowerbirds of Queensland and New South Wales, Australia. Unlike a number of other male birds who use plumage to attract mates, the male bowerbird, a creature the size of a robin, tries to win the heart of the female by offering a house—correct that: a castle. These bowers can be up to five feet high and six feet long, and the males brighten the places with blue and yellow objects they find in nature, or, as it happens, in the researchers' cabins. "They will break in and steal ballpoint pens and clothespins," says Gail. "It's probably one of the few places where if you discover a pen missing you know where to find it." Conversely, if the researchers are out in the field and need a pen, they will first scan one of the bowers to see if one is handy.

While the male bowerbird is building his honeymoon

house, the female sits in a tree looking over the work with an appraising eye. If it meets whatever criteria a female bower-bird uses to judge the relative merits of bowers, she will then fly down and enter it while he goes into his courtship dance. The economics behind this elaborate ritual would be instantly understandable to any gold-digging man or woman.

The evolutionary pressures that favored male birds who built bigger and better bowers involve the issues of honesty and deception. According to behavioral ecologists, male displays like gaudy plumage, elaborate birdsongs, and expensive gifts all serve as signals through which male suitors convey the message that they are robust and successful with good genes to pass on. All such signals are designed to protect females from being seduced by counterfeiters and poseurs who might pretend to be healthy and aggressive but really are not.

The idea, called the "handicap principle" in behavioral ecology, is to make the signal so expensive that it cannot be faked—"like taking your date to St. Bart's for the weekend on your private jet," as one scientist put it. The peacock's over-the-top tail tells females that the male is strong enough to hold it up, while the bigger the bower, well, the bigger the . . .

At the 1997 International Behavioral Ecology Congress meetings in Asilomar on California's Monterey Peninsula, a good deal of discussion focused on attempts to model the balance of cheating and selfishness versus honesty and cooperation in various animal societies. One such simulation developed by mathematician and theoretical biologist Thomas Sherratt showed that if an animal follows simple rules such as limiting their investment in strangers, reciprocating kindnesses, and paying less attention to those who make increasing numbers of mistakes (and/or cheat), cooperation tends to beget more cooperation. As Sherratt noted in a subsequent communication, "Reciprocation is a powerful stabilizing force, promoting cooperation, but punishing cheats and short-changers." Cheaters might do well for a generation or two, but over time cooperation wins out—a lesson some of the more corrupt human societies might note.

Deception has its positive aspects for the group at large. Some convenient fictions, for instance, when entered into collectively, help preserve comity and reduce violence. After suffering through a year of preoccupation with the sexual adventures of President Clinton, most Americans yearn again for the days of the Kennedy administration, when everybody in Washington knew about JFK's sexual antics but pretended that they were not happening. As we enter a new administration, maybe we should take a page from baboons, who have figured out artful ways to use collective lies to defuse tense situations.

Hofer and East tell the story of how an ambitious young male olive baboon found himself in a situation ripe for conflict. The baboon was in a food pit at the Gombe Stream Reserve in Tanzania at a point when the dominant male baboon unexpectedly arrived at the edge overhead. The alpha male started lip-smacking—in baboon-speak a signal to get out of the pit or face a fight. This put the junior baboon in a bind. He could either rise to the challenge and risk a savage fight whose outcome was by no means certain, or he could slink away, losing face and status in the eyes of the dominant male as well as the other baboons watching the situation unfold. Instead he chose a third path, and embarked on a challenge to a far-off male baboon in no way involved in the situation.

After a moment, the alpha joined in the charade, redirecting his own challenge to the distant baboon, who, minding his own business far away from the food pit, was in all probability completely bewildered by this sudden animosity coming from two of the more dangerous males in the troop. Now joined in common cause with the alpha male, the junior baboon in the food pit suddenly charged the more distant male, who took off for safety. Net result: The alpha male gets the food pit to himself, both baboons avoid a fight, and the young challenger saves face.

Was this just a blind reaction, or were these baboons actually aware of the risks, and were they strategizing to avoid them? Robert Seyfarth and his wife, Dorothy Cheney, are two University of Pennsylvania wildlife biologists who have conducted imaginative studies of animal awareness with vervet monkeys and baboons in Africa by watching their reactions to tape recordings of the sounds of various predators and various forms of alarm calls. They have also conducted extensive field observation of the animals' social interactions.

One of their more interesting findings is evidence that baboons are aware of their status in the hierarchy and react accordingly. (As Seyfarth notes, it is one thing to look around you and know who is boss. It is quite another to see a hierarchy and also be aware of your place in it.) The basis for Seyfarth and Cheney's reasoning was their observation that when a dominant female baboon approaches two other females lower down in the hierarchy, the most junior of the two baboons is the likeliest to move off. The more senior of the two being approached might be saying to herself, "Uh-oh, Big Bertha's approaching, but I can hang around because Junior here is going to acknowledge her status by going away." With up to thirty-nine females in the groups he was studying, Seyfarth says that the baboons could not simply be memorizing various combinations. Here is a case where self-awareness is actually the simpler explanation for a particular behavior.

If an animal is aware of its status, then it is possible, if not indeed likely, that the animal is aware when it deceives another animal. It is also likely that the animal is aware that it is taking a risk when it deceives another member of its group. Such awareness may well keep deception in check inside and among various social groups.

On the other hand, while awareness may be involved in deception, we must be careful not to ascribe higher consciousness to a piece of treachery that might be the product of simple reinforcement. For instance, primate specialist Frans de Waal observed a chimpanzee who was left with a temporary limp after he lost a fight to a more dominant male. The loser quickly figured out that the stronger male would not beat him up when he was limping, and so, even after his leg had completely healed, he resorted to limping whenever he saw his antagonist. While affecting a limp requires no more cleverness than making a simple association between a behavior and the avoidance of pain, other types of deception cannot be explained so simply.

Indeed, de Waal was at first unprepared for the amount and sophistication of deceit in chimp life when he first began to observe the animals. In his 1996 book *Good Natured: The Origins of Right and Wrong in Humans and Other Animals*, he wrote, "With years of macaque watching behind me before I was introduced to chimpanzees, I was totally unprepared for the finesse with which these apes con each other. I saw them wipe undesirable expressions off their face, hide compromising body parts behind their hands, and act totally blind and deaf when another tested their nerves with a noisy intimidation display."

Take, for instance, the case of a double deception witnessed in Tanzania's Gombe Stream Reserve, the field research station pioneered by Jane Goodall. Scattered throughout the area were some feeding stations that contained treats. One of the Gombe chimps had the luck to be alone right next to a feeding box when it was opened by remote control. Noticing that a more dominant chimp had come into the area, the first chimp quickly closed the box and moved nonchalantly away until the senior chimp had moved on. Once that chimp was out of sight, the first chimp returned to the box and opened it to claim the food. The more dominant chimp, however, had suspected that something was amiss and hidden himself just out of sight. The first chimp's elaborate attempts to conceal the booty were nullified as the dominant chimp returned to snatch the bananas. Clearly the dominant chimp had not secured his position by brawn alone.

Behavioral ecologists spend a lot of time trying to sort out the meaning of various deceptions. The first question is whether the animal controls the signals it gives. Clearly a tropical fish whose coloration makes it look larger than it is does not control such a signal.

On the other hand, controlling body language seems to require some awareness. But how much? In the early 1970s, I owned a Maine coon cat named Zephyr. My wife and I then lived in central New Hampshire, and it could get bitterly cold in the winter. When Zephyr wanted to come in, he would sit on the railing of the deck outside the kitchen window in view of those inside and meow. This was usually quite effective, but sometimes when there was a wind, we could not hear his cries. On one such night I was at the sink when something caught my eye. It was Zephyr. Instead of sitting on the rail, he was on the deck and jumping straight up into the air. I am convinced that he had decided that if a meowing stationary cat couldn't catch my attention, perhaps a moving object (himself) might.

Because we know that we are blessed with awareness and intelligence, and have used these terms to describe what we do, think, and feel, we tend to assume that human behaviors reflect consciousness when in fact they are automatic. When the tired commuter slouching towards the 5:28 to Greenwich passes a pretty woman and straightens up and pulls in his stomach, is he really conscious of what he is doing? At some level perhaps, but it could also be a type of sleepwalking—an unconscious reaction produced by ancient biological imperatives. Looked at by a skeptical observer, much of our behavior would not satisfy even our own definition of consciousness. As one psychologist put it, "Keep in mind that from a maid's perspective, all hotel guests are robots," in the sense that they mess up rooms in nearly identical ways.

In fact, there are plenty of human behaviors that start out as conscious but then, through repetition, become automatic. This makes sense, of course. Storing an insight or new behavior as a set of compiled rules frees up precious mental space for new challenges. Driving a car is a good example. Much of the time we are on automatic pilot. We awaken only when our senses alert us to danger or something out of the ordinary.

When the junior chimp affected a nonchalant demeanor in the presence of the bananas, he was overruling a natural urge to express excitement at the discovery of his windfall. That in turn suggests that the chimp at some level realized (even if only as a memory of previous bitter disappointments) that his natural reaction would attract the big male.

By contrast, the dominant male's deception was a work of art. He read the situation and smelled a rat. Instead of blustering over to the junior chimp, he continued on his way and hid out to see how the situation developed. Of course, it is possible that the observing scientist mistook mere coincidence—the dominant male just appeared to be hiding and only happened to look back—for something purposeful. Was the dominant chimp aware of his effect on the subordinate male and adjusting his behavior accordingly to achieve a goal? The next chapter will delve into this question.

I THINK THAT YOU THINK THAT I THINK THAT YOU THINK . . .

Mind Reading and Mental Chess

I N the previous chapter, we saw examples of animals attempting to second-guess their peers as well as the humans around them. Even if Koko's and Chantek's attempts
to shift blame are risible, these efforts may reveal fumbling
attempts to plant false ideas in their keepers' heads, implying that the apes realize that others can have a false idea.
While these examples might be dismissed as ambiguous in
the hard-nosed world of controlled experiment, other scientists have subsequently discovered that gorillas, chimps,
bonobos, orangutans, and very likely dolphins and orcas
share a degree of consciousness with humans that permits
them to guess the mental state of others around them.

Most zookeepers would say they already knew this, but it
is no small matter to prove this concept of consciousness to a
skeptical audience because of what it implies about the mental world of these higher mammals. A person or an animal
has a tremendous advantage if able to reason about un-
observable mental states in a rival—to know what he or she
thinks, and then to judge how that rival might be encouraged to think something that works to one's own benefit.
If you can know that one of your fellows is wrong about
something—e.g., that the keeper thinks that you have not

yet gotten an orange—you can capitalize on his or her false beliefs.

The key to all of this is understanding that another creature can be wrong about something. Most animals recognize at some level whether their peers are angry, lustful, or in any number of basic states that can be read from body language and posture. But what about being mistaken? That requires projecting oneself into another creature's head; reading its mind. Cognitive scientists call this the ability to have a "theory of mind." If you are capable of having a "theory of mind," then you have a powerful tool for manipulating your peers to your advantage.

Cognitive scientists have speculated about these questions for many years. In 1978, David Premack, who abandoned exploring chimp language abilities for work in animal cognition, published a paper entitled "Do Chimpanzees Have a Theory of Mind?" in which he speculated that one aspect of consciousness is the awareness that others can be mistaken or have knowledge different from one's own. In response to this, Daniel Dennett wrote a letter proposing that one way to test this idea empirically would be to devise a Punch-and-Judy show suitable for both animals and humans. As earlier generations know, this involves a little pantomime in which the viewing children know that Judy is in a box, and also know that Punch does not know that.

Premack took this simple suggestion and devised an ingenious test to explore at what point children begin to understand the mental states of others. The test he devised was simplicity itself. A child would be shown a little play in which a girl named Sally enters a room and puts a marble in a bag and then leaves the room. Before Sally returns, another little girl, Ann, enters the room and takes the marble from the bag, putting it in a box. Sally returns, and the child is asked to point to where she will look for the marble.

It turns out that three-year-old children watching the performance tend to point to the box where the marble is. By the time most children are four, however, they know that Sally has the false belief that the marble is still in the bag, and that she will look for it there. Four-year-olds, in short, are able to reason about the nonobservable mental states of other people.

This test gave comparative psychologists a way to perform similar studies of apes and other animals. It was adapted so that comprehension of language was not necessary, and also to make the charade interesting to a chimp, orangutan, or monkey. In one version run by Daniel Povinelli, then a psychologist at the University of Southwestern Louisiana, chimps had to choose which of two humans would be better at helping them find some hidden food. The chimps themselves could not see where the food was hidden, but they could see that one of two humans in the area had a full view of the process as the food was being hidden. The chimps were then asked to choose a human helper to get their food. They overwhelmingly chose the person who actually saw the procedure, suggesting that they realized that the other human did not know where the food was hidden and thus would be less helpful.

Povinelli also tried to determine whether smaller-brained primates also had insight into the knowledge of others. It turns out that just as there seems to be a threshold of consciousness between three- and four-year-old children, a divide also separates the great apes from monkeys. When Povinelli duplicated his experiment with rhesus macaques, Asian monkeys best known for their begging at Indian and Nepalese temples, the primates chose human helpers randomly, showing no preference for those who actually knew where the food was hidden.

Another scientist, Juan Gomez, tried a similar experiment with an orangutan at the Madrid Zoo. This version was a little closer to the original Sally/Ann test. In this primate play, the orangutan watched as a provider put goodies in one box and a key in another. Then he would leave and return. When he came back in, the orang would point to the box with the food and the keeper would get the key, open the box, and give the orang a treat. Every so often, however, the keeper would leave and another person would come in and move the key. On these occasions, instead of pointing to the box where the food was, the orang would point to the box where the key was, seemingly aware that the keeper was misinformed about the location of the key.

To control against coincidence, Gomez also staged trials in which the keeper returned and moved the key himself before leaving. In these instances, the orang did not point to where the key was, presumably because he remembered that the keeper himself had moved it and therefore knew where it was.

Working with Rob Shumaker, another Think Tank scientist named Dan Sillito also did a variation of the Sally/Ann experiment. "We reduced the problem to the question of whether an ape can know that what you know is different than what someone else knows," says Shumaker. As a first step one experimenter would hide food under one of two cups in sight of the apes and then leave. Then they determined whether the apes could direct a new person to the place where the food was hidden. The answer of course was yes. They then repeated the experiment with some twists. In one iteration, two people would come in after the food was hidden, but one of the two would have a basket over his head that obscured his vision. In this case, the apes would choose the person who did not have the basket over his head to get the food. In another version, one person would wander in with a basket over his head. In this case, the orang would remove the basket and direct him to the food.

The final trial involved a case where Rob was in a situation where he could see a treat like cookies but couldn't reach them. Moreover, the orangutans had a number of tools in their part of the cage that would enable Rob to reach the cookies. Rob would ask the orangutans for help (which of course assumes that they understand English), but he would not ask for a tool. On almost every occasion, they gave Rob the appropriate tool.

On one occasion, Indah, one of the female orangutans, gave Rob a tool that would have been a sufficient retrieval device for an orangutan because they have such long arms, but which was not long enough for Rob. Seeing that he could not reach with the tool she had given him, Indah went and found a longer stick and handed it to Rob. On another occasion, the appropriate tool was in Rob's part of the cage, but he did not know it. This time Indah reached through the cage and grabbed the tool, pulled it into her part of the cage, and then pushed it through to Rob.

Some of the experiments probing the degree to which animals are aware of the mental states of those around them approach the farcical. In one case, the experiment called for chimps to choose between two people to deliver cups of juice as a treat. One person, acting the role of a klutz, accidentally spills the juice in the cup on the way in. The other person, playing the role of an evil keeper, would deliberately spill the juice in a cup before delivering it. When asked to choose who they want to deliver juice, the chimps decidedly favored the klutz, apparently making the commonsensical judgment that a well-intentioned clumsy person was vastly preferable to a handler with sinister intent.

Researchers tried a similar experiment with capuchin monkeys, only this time the sinister helper would actually eat the treats in full view of the monkeys rather than deliver them. Even after 150 repetitions, the poor monkeys pathetically continued to put their trust in the now bloated evil human helper.

Perhaps the most compelling indication that a chimp can be aware of the state of knowledge of another chimp has yet to be published. Working with a chimp colony at Ohio State

University, Sally Boysen adapted an experiment that Seyfarth and Cheney had earlier conducted with vervet monkeys. The experiment tried to determine whether a mother
vervet monkey responded differently if both she and her infant could see a threat than when only she could see the
threat and it was clear to her that the infant could not. In
essence they were trying to determine whether the mother
made distinctions concerning the state of knowledge of the
infant.

Boysen decided to do this with chimps. In her version, she
would hide in a cage within sight of Darryl, an adult male in
an adjoining cage, but out of sight of Kermit, a young chimp
who was let into the cage. Darryl made no attempt to alert
Kermit to Sally's presence if Sally was holding food or any
number of inoffensive objects, but on those occasions when
she was armed with a tranquilizer gun, Darryl would launch
into a dramatic display as soon as Kermit was let into the
cage. For his part, Kermit would stop dead in his tracks and
turn around and exit without even investigating what Darryl was making a fuss about. This was duplicated with several other chimps playing Darryl's role, and according to
Sally, they only showed stress or fear when Sally had the
gun and when another chimp was approaching the blind
where Sally was hidden.

If animals have some degree of self-awareness and an
ability to speculate on the knowledge of others, a natural
question follows: What role does this ability play in an animal's daily life? Humans (sometimes) can overrule their appetites when the application of intelligence leads them in
another direction. Can animals do that too? It turns out that
one of Boysen's more provocative experiments probed the
degree to which a chimp's ability to reason is subservient to
its appetites.

This experiment involved two chimps: Sheba, a female at
Ohio State University, and Sarah, the chimp who had been

the subject of David Premack's early language experiments. The experiment centered on a game in which Sheba would be shown two dishes filled with different amounts of treats. The first dish she pointed to would be given to Sarah, meaning that Sheba had to think smaller to get larger. When she could actually see the treats, Sheba invariably pointed to the larger amount, only to see them given to Sarah. However, when tokens were substituted for real food (which was then given to the chimps after Sheba pointed), Sheba instantly realized that pointing to the smaller amount would get her the larger amount. It would seem that when she was tempted by real food, Sheba's appetites persistently overruled her ability to reason. When temptation was removed, Sheba could bring her cognitive abilities to bear on the problem and achieve her desired, albeit selfish, result. The split-second difference between seeing the treat and reacting and seeing the symbol and assessing it was enough to reroute the sensory information in a way that would allow her analytical abilities to come into play.

Psychologists have tried similar games conducted with children of different ages, autistic children, and children with Down's syndrome. Four-year-old children and children with Down's syndrome realize that if they point to a smaller amount of food, they will be rewarded with more. Three-year-old and autistic children don't. This suggests that sometime during human maturation, a child's cognitive abilities develop to the point where they realize that only by overruling their urge to pick the larger amount of treats will they actually get what they want. The evidence also suggests that Sheba and other chimps are right on the cusp of that threshold. "In the course of an afternoon, we could toggle between Sheba reacting like a three-year-old and a four-year-old simply by switching what she was looking at," says Boysen.

Marc Hauser of Harvard has also delved into this question of the body overruling the mind. With regard to the

Sally/Ann test, for instance, he points out that even though younger children may point to the box where the marble is, they often look at the box where Sally will look, suggesting that they know that is where Sally thinks the marble is, but persevere in pointing to the bag holding the marbles. According to Hauser, as in the case with Sheba, this is because reality is more powerful than thought at that point in their little lives.

British psychologist Andrew Whiten of the University of St. Andrews in Scotland has devoted his career to the study of deception and awareness in the animal kingdom. He says that the dramatic differences these studies reveal support the notion of a "mental rubicon—not the familiar one with humans on one side and everyone else on the other, but with man and at least the apes on the same side."

As with any exploration of higher mental abilities, nothing is as clear as we would like. Because it took a large number of trial runs before Povinelli had his experiment running smoothly, he has backed away from assertions that chimps have a theory of mind. On the other side of the "mental rubicon," tamarind monkeys may have some awareness of the mental states of their peers, according to Marc Hauser. But these asterisks aside, Whiten's enthusiasm for a mental rubicon is probably warranted, and the chain of experimentation that Premack and Dennett provoked is one of the livelier developments in the study of animal awareness in recent decades.

Those animals that turn out to be on our side of Whiten's mental rubicon are endowed with at least one of the fundamental abilities necessary for that fuzzy concept that we call intelligence. We may not know what intelligence is—just one ability or an ensemble of skills that we lump together under one word—but we can describe what it does.

Intelligence enables animals to explore alternatives in their minds before taking risks in the real world. A lot of this involves determining the intentions and anticipating the actions of those around them. As Andrew Whiten puts it, "In chess, or even Ping-Pong, a good player is able to second-guess the intentions of others and predict what they might do next." It's one thing to be able to do this on the basis of body language, but quite another to be able to probe beyond body language to the invisible world of thought in another's cranium.

The issue of reading another's intentions also gives some clue as to one of the evolutionary pressures that might have produced such abilities. While monkeys must search for clues, a chimp can decide who he trusts. Says Daniel Povinelli, "It's like the difference between a random search and following a highway. It makes you a more efficient learner and reduces your search space by telling you which stimuli are important."

It is probably no accident that animals with the most developed ability to read the intentions of others, who use that knowledge to deceive, are also those animals that live in large social groups with complex and changeable social interactions. Indeed, British anthropologist Robin Dunbar has shown that the bigger and more complex the social group, the bigger an animal's neocortex, the part of the brain where we do our scheming and plan our deceptions.

Why hasn't every animal developed such abilities? For one thing, the bigger brain required by mental chess diverts blood from the muscles. Nature has the leisure of millions of years to run her experiments, and the relationship between complex social networks and bigger brains suggests in a crude way that the security provided by a big group justifies the diversion of blood to the brain. We can say that nature has maximized this type of awareness in those species where advantage conveyed by knowledge of what is going on in the heads of others outweighs the risk of diminished physical response.

There is no question that as a species we have been optimized for such activities. There is also no question that the price of our blood-guzzling brain has diminished our physical robustness. All the apes save gibbons (which weigh just a few pounds) have several times our strength. Yet while orangutans and dolphins have a foothold on our side of the mental rubicon, we certainly command the heights. In virtually all studies of animal intelligence and language skills, performance plummets as more elements are added to a task and the longer an experiment requires that an animal remember those elements.

Zookeepers cheerfully exploit this advantage in staging their own tricks to get animals to do what they want. Violet Sunde, a longtime gorilla keeper at the Woodland Park Zoo, used a particularly devious ruse to get Kiki, an adult male, to cooperate with *National Geographic* filmmakers who were visiting the zoo. Kiki had a habit of climbing a tree in the exhibit, which made for a picturesque sight, but suddenly camera-shy while the filmmakers were there, he refused all entreaties to climb the tree.

After a few days, the *National Geographic* people began wondering if the gorilla would ever climb the tree. Violet decided to intervene. The exhibit area was monitored by cameras that fed images to a screen on a desk in the holding area. When they were in the holding area, the gorillas could see the monitor. Violet and the other keepers would leave the monitor running even when they weren't there so that the gorillas could see what was going on in the exhibit. Using remote controls one morning, Violet turned the camera towards the tree in the center of the exhibit. Then one of the more agile zoo workers climbed the tree, wedging various treats into the branches. A little later, Kiki was let out of his cage. Rolling cameras notwithstanding, he made a beeline for the tree and began climbing.

One could say that Violet outwitted Kiki, but look at the assumptions Violet made in confecting her ploy. She had to assume that Kiki would make the association between the image on a small screen and a reality that was out of sight, in itself an ability that has been the subject of experiment and debate. Violet's attempt to fool Kiki was predicated on the assumption that he possessed to some degree the higher mental abilities we would use to make the connection between images on the screen and real events occurring somewhere else but connected to his world. In short, she assumed Kiki was aware.

With this in mind, let's revisit some of the examples of deception cited earlier. Once an animal demonstrates some capacity for awareness, it is highly likely that this capacity bears on his critical interactions with the important figures in his or her life, particularly since the arms race of the intellect that produced awareness in the first place probably grew out of such interactions.

Thus, it is not stretching things to assume the chimp that affected a limp to avoid a beating was exploiting his awareness of what I guess might be called a primatarian streak in the higher-ranking male he feared. Similarly, in all likelihood the two male chimps involved in the double deception at the Gombe Stream Reserve probably were trying, as Winston Churchill once put it, to protect the truth "with a bodyguard of lies."

So too was a captive four-year-old gorilla named Zuri when a treat was tossed to him at the Woodland Park Zoo. When he noticed that Jumoke, a big adult female, was charging over to take the food, he studiously pretended to look for it in the grass of the outdoor exhibit. Jumoke started looking also, and then gave up and went away. As soon as she was out of sight, Zuri abandoned his phony search through the grass and made a beeline in the other direction to the food.

Think again about the gorillas who used ruses to lure people within reach and then tried to poke them or grab their shirt. Through their body posture and gestures they were attempting to plant a false belief in the human's mind—e.g., "I can't reach you" or "Please come here, I've got something to show you"—with a plan to manipulate that false belief in

order to play a trick or get revenge. This is the very essence
of what the Sally/Ann test seeks to uncover.

I would not be surprised if subsequent testing of elephants, parrots, dolphins, orcas, and some other higher
mammals also showed that they could be aware of the mental states of other creatures. And maybe even some unlikely
animals will eventually prove to have similar capacities. Remember my late cat Zephyr, who jumped up and down to
get my attention. The arguments presented in this chapter
would suggest that a cat should not be able to interpret the
interior mental state of a human. But then again, cats are always doing what they shouldn't be able to do.

Indeed, just when you are prepared to consign cats to one
of the lower circles of mammalian intellect, they confound
things by doing something relatively sophisticated. At a conference on animal cognition I attended in Strasbourg, France,
a major session was devoted to exploring the question of
teaching in the animal kingdom. The way cognitive scientists define it, teaching is a sophisticated cultural behavior
because it is distinguished by self-sacrifice on the part of the
teacher (for instance, it is not teaching if an animal simply allows its young to watch while it performs some task for its
own benefit), and by other indications that the teacher is attempting to impart a skill.

During this session, mention was made of one example,
described by Christophe Boesch, a Swiss scientist, of a chimp
teaching its young to crack open nuts using a rock in the Tai
Forest in West Africa. Another researcher brought up a case
in which wild orcas were observed circling prey and seemingly guiding their young in hunting techniques. There was
some discussion of these two admittedly skimpy examples,
and there was some scratching of heads about the paucity of
evidence for teaching in some of the smartest creatures in the
animal kingdom.

Then someone mischievously raised the issue of tiger
teaching. It turns out that tigers come closer to meeting all of

the criteria scientists have established for pedagogy than almost any other animal. Tiger cubs, for instance, have to learn
when to kill prey by seizing its throat and when to kill it by
biting the back of its neck. (For those would-be tigers among
the readers, when the prey animal has horns, the preferred
method for the kill is to seize the animal by the throat rather
than the back of the neck, in order to minimize the risk of a
debilitating injury.) Does this mean that cats are smart? Or
does an animal not necessarily have to be smart to meet the
criteria for teaching? Heeding the words of Donald Griffin,
who argues that awareness might be very generously spread
through the animal kingdom, I try to keep an open mind.

It is easy to present a cat, dog, or even a dolphin, chimp,
or orangutan with some stimulus that produces an instinc-
tive response. Do it enough and the animals begin to look
like windup toys, with the differences between animal and
human looming larger and larger. But the work by Shumaker, Sillito, Hauser, Boysen, Whiten, Byrne, and many
others suggests that behind and alongside these hardwired
responses exists some measure of awareness and consciousness in a number of species. Whether a person sees the animal kingdom as populated by automatons or enlivened with
sentience thus depends more on how we look at other
creatures than on what is inside the heads of the animals
themselves.

THE PIG WHO
RAN TO WORK

*Cooperation in Work,
Conflict, and Healing*

WHY would an animal want to cooperate with a human? The behaviorist would say that animals cooperate when, through reinforcement, they learn it is in their interest to cooperate. This is true as far as it goes, but I don't think it goes far enough, if we remember that there are rewards and there are rewards. Certainly with humans, the intangible reinforcement that comes from self-respect, dignity, and accomplishment can be far more motivating than material rewards. Is it also possible that it is important for an animal to feel that it has some purpose, that for social animals that purpose involves meaningful interactions with others, and that the self-respect gained from cooperation might be more important than getting a cracker?

Zookeepers recognize this. Those who work with animals have been doing their best to try to understand the intangible as well as material needs of different animals in order to relieve the boredom and purposelessness of captivity. Through enrichment programs, a cheetah might still feel the thrill of the hunt by chasing after make-believe prey in a specially constructed race at the San Diego Zoo. In Scotland, tigers must run up a twenty-foot-high pole to get to food at certain times. As Karen Pryor points out, zoological workfare like this makes for happier tigers. It allows them some sense of the hunt, which has to be part of the enjoyment of a meal.

Zoo enrichment programs give back to animals some element of control, and with it some measure of dignity. Much of the mischief, tricks, and escape attempts perpetrated by captive animals seems to be motivated by a desire to assert independence and control. "True enrichment," says Rob Shumaker, "is not about balls and poles, but about what we all want, which is control over our surroundings."

This brings us back to cooperation. For the most part, animals do not accept the terms of captivity passively, and cooperation is another way an animal can assert some control and find some meaning in the terms of its confinement. Indeed, according to Gail Laule, Orky the killer whale turned the tables and made cooperation a reward for his trainers when they treated him with respect. This is true for many proud animals. Chris Wilgenkamp, the elephant keeper at the Bronx Zoo, says that the elephants love interactive games with the keepers because the pachyderms control the interactions.

Enriching the animals' lives also makes them more cooperative during veterinary examinations. By turning an examination into a fun game with rewards, Wilgenkamp and his fellow keepers have taught some of the young tigers at the Bronx Zoo to come running up to the wire mesh and show zoo doctors their paws and bellies. Sometimes cooperation extends to the point of reminding zookeepers how to do their job. Loraine Hershonik, a senior lion keeper at the Bronx Zoo, says that on one occasion an old female lion named Kathrin noticed that a gate to an outer cage was nearly unlocked. "She roared," says Loraine. "It was almost as though she was reminding me: 'This is not the routine.' "

A number of captive animals actively mimic the keepers' daily routines. Ivan, for instance, will clean his cage if you give him a towel. When I first arrived at Tanjan Puting, the orangutan rehabilitation/research station run by Birute Gal-dikas in Borneo, I passed a young female orangutan furi-ously scrubbing a piece of clothing in imitation of the local girls who were washing clothes on the deck. And when keepers at the Topeka Zoo found a male orangutan named Jonathan outside the holding area after one of his many es-cape attempts, the big male was mopping the floor with a squeegee, causing one of the watching keepers to suggest, only half facetiously, that they let him finish before getting him back into his cage.

Cooperation is most interesting when it comes from ani-mals from which it is most unlikely—as, for instance, in the case of the big cats. In terms of intelligence, cats, big and small, are an enigma. They are adaptable and flexible, but by the traditional formulas by which scientists determine potential for intelligence, such as brain-to-body weight, they should have nowhere near the capacity for learning or awareness of monkeys and apes. As Reduoun Bshary, a be-havioral ecologist, put it to me when we were discussing leopards, "Why should a leopard have to be smart? It's al-ready perfect."

But (it can't be said too many times) cats delight in doing what they are not supposed to do—or be able to do. Anyone who knows cats knows that they are in some ways very smart, in some ways dumb or at least stubborn. What is be-yond dispute is that any cooperation one receives from a cat, big or small, is a grand gesture on the part of the cat.

Rick Glassey, an animal trainer, has since 1974 made his career based on his ability to get big cats to cooperate in the making of feature films. He got into the work because of his love of the big cats, and his group of tigers, lions, moun-tain lions, leopards, and panthers have acted in dozens of movies, including such recent films as *Dr. Dolittle* and *Mighty Joe Young*. He says that the key to establishing a relationship with the big cats is to get them to recognize that you are being fair. Indeed, because they are so sensitive to unintentional slights, Rick says that he won't work with them on days when he is in a bad mood. (The notion that big cats can read minds was expressed by a number of keepers as well. Maybe they do have a theory of mind.)

All Glassey's cats were born in captivity, and he established a strong relationship with them early on, but he insists that the relationship is not built on fear. "People come here [to his base facility] all the time," says Glassey, "and want to go out with the cats. They'll say, 'I'm not afraid of them,' but, you know, I'm not sure the cats really care whether you are afraid of them."

Being fair means respecting the cats' desire to stretch their legs and pretend that they are big cats out in the wild, and so Glassey takes them out for regular walks through the high desert surrounding his enclosure in Nevada. I accompanied Rick on a couple of these walks with, in turn, an adolescent tiger, a young lion, and an eighteen-year-old tiger named Jake who still works in movies. Even though they weren't on a leash, the cats all respected the terms of the walk, and neither pounced on me (something Rick assured me that they would do if he weren't around) nor made any attempt to escape.

As we walked through the desert, there was a comical scene when we approached a road leading towards a wildlife park. Seamus, the nearly full-grown male tiger, had been having a great time going off in the brush and investigating. At one point, when a car approached, he crouched down behind a clump of sagebrush, very much into being a stealthy tiger. Then he got up and resumed his walk beside Rick while the car drove on towards the park, its passengers looking forward to seeing bear and elk and blithely unaware that they had just passed a tiger roaming free not one hundred feet from their car.

When the big cats are let out, one thing they do is run up to greet the other cats in their cages, even though most of these different species would be mortal enemies if they ever encountered each other in the wild. Thus, when Mufasa, the thirteen-month-old lion, is let out, he promptly runs up to the adjoining cage that holds the young tigers, and in a friendly gesture rubs his side up against the bars. "They are best buddies so long as they are in the cage," says Rick, adding that the tigers play too rough for the young lion.

Mufasa then follows us into the brush and soon takes off by himself for a little exploring. Rick shows not the slightest concern. "I'm his pride," he explains, "so he never goes very far away. Most of the time when we go out, I just walk and he will follow." After a while, Rick decides it's time to bring the young lion back in. Peering off into the brush and seeing nothing, I ask how he knows where he is. "Just look at the other cats," he says. Sure enough, they are all at the backs of their cages staring enviously off at a point some fifty yards behind the enclosure.

Glassey says that the big cats know very well when they are doing something bad. After he lets Jake out, the five-hundred-pound tiger first greets the other cats, and then goes over to a patch of lawn where Rick has a sprinkler going. One of Jake's favorite games is to try to bite the moving stream of water. (This is a lot easier on Rick's wallet than one of the tiger's amusements when he was a rambunctious two-year-old living on a farm in New Jersey. Back then he would run up to farm tractors and puncture the five-hundred-dollar tires with one huge bite). After a moment of playing with the spray, Jake moves to bite the hose. In a stern voice, Rick instantly says, "Jake no!" Jake looks up, all innocence, and chuffs (a tiger grunt that means everything is OK) as though to say, "I wasn't going to do it." Earlier in his career, Glassey worked at Marine World in California, and he remembers one tiger who would knock down trees that Rick had planted. If the tiger saw Rick coming after he had knocked down a tree, he would run away and cover his eyes with his paws. Perhaps, like a two-year-old human toddler, the tiger thought he would be invisible if he covered his eyes.

Despite his remarkable rapport with the big cats, Glassey recognizes that there are limits to the cooperation he can expect from his feline actors. Film directors, says Glassey, tend to look at big cats as if they were furry actors, but, he notes, "if you can get them to run, jump, stop on a mark, and snarl, you're getting a lot."

If creatures less perfect than big cats have to be smarter in certain ways, they still base their willingness to cooperate with humans on the fundamental precepts of trust, understanding, and respect. Gail Laule remembers a number of situations when Orky made clear his feelings that he thought such respect had not been forthcoming. On one such occasion, Orky and a trainer were practicing a routine at Marineland in which the trainer would ride around the tank on the orca's back. Spotting a pretty girl, the trainer had Orky pull over. The trainer stepped off and started flirting with the girl while Orky cooled his flukes by the side of the tank. When the trainer finally turned around to step back onto the orca, he found himself stepping not onto the giant dolphin's back, but into his wide-open mouth.

It turned out that Orky had used this device more than once to let trainers know that his cooperation came in return for respect. Gail says that there were stages to Orky's anger. First his eyes would turn red. Then, she says, he might spit a bit of water at you. If you still did not get the message, Gail says he would set his chin on the edge of the platform from which the trainers gave commands and pump his tail, Gail notes, "as though to say, 'I'm an orca and I could get you if I wanted to.' " Or, if it was a day when a trainer was on his back, he would take the trainer out into the middle of the pool and not bring him or her back. To express extreme irritation, he would spit more violently. Despite the escalating threats, Gail says that he was always in control of his temper, and that after a tantrum he would swim up and let someone give him a back rub.

In one instance, a particularly macho trainer failed to respond to any of Orky's increasingly urgent signals that he did not like the way the trainer was treating him. Pressed beyond his patience, Orky chose to make his point during a show. In a flash, says Gail, he erupted from the water in an unscheduled improvisation on the routine, grabbed the offending trainer in his mouth, and then let him go. Even though the trainer did not get the message, the rest of the Marineland staff did, and the man was reassigned to work with other animals.

The Marineland staff not only needed to respect Orky to get him to work, they also needed his cooperation if they were to work with his mate, Corky. At least in captivity, orcas seem to have very old-world ideas about gender relationships, which is to say that Orky controlled Corky. When Corky was not cooperating in a routine, the trainers needed to figure out whether this was because the female orca was being moody, or whether she was following instructions from Orky. For instance, on some days Orky would be performing perfectly, but Corky would lurk sullenly out of the action. On these occasions, the trainers would refuse to give Orky the customary reward. If he was not controlling her, he would get angry at the injustice. If, however, Orky had been directing Corky to stall, he would then let her rejoin the session.

"We really needed to figure out what was going on," says Gail, "so we worked out a compromise and made a deal." If Corky was misbehaving because of Orky's influence, he would go get her and in return be rewarded with fish and expressions of gratitude. Gail says that Orky's control over Corky was his way of expressing dominance—not so much over Corky as over the humans in the unnatural situation of captivity. One note: Orky's old-world attitudes towards females also had their chivalrous side. Gail says that when

Corky was pregnant, he became a marshmallow and she could "walk all over him."

Laule also uses operant conditioning in her training and counseling work. Although she uses behaviorist techniques, when Gail talks she sounds like anything but a behaviorist. "I use operant conditioning," she said to me, "but I recognize its limits. For instance, it's much easier to work with a dolphin if you assume that it is intelligent." Then she went on: "That was certainly the case with Orky. Of all the animals I've worked with, Orky was the most intelligent"—here she paused a bit as though looking for a word—"and deep. He would sit back and assess a situation and then do something based on the judgments and assessments he made."

Gail's most memorable example of this once again involved a baby. Corky gave birth to a baby orca fathered by Orky in the late 1970s. The baby did not thrive after being born, and after about two weeks, the keepers decided to take the little killer whale out of the tank for emergency care and feeding. To avoid disturbing Orky and Corky, the keepers decided not to lower the water level in the tank but rather to get the baby onto a stretcher in the water. They successfully maneuvered the baby onto the stretcher and completed the feeding, but when it came time to return the infant to the tank, things began to go awry. According to Tim Desmond, who was present during this operation, the boom operator who controlled the movements of the stretcher did not have a clear line of sight to the tank, and he halted the stretcher when it was still a few feet above the water. At the same time, the baby began throwing up, through both its mouth and its blow hole. The danger was that it would aspirate some vomit, which could bring about a fatal case of pneumonia. It was, in Tim's words, a "desperate situation" because the keepers treading water in the tank could not reach the baby dangling above.

Orky had been watching the procedure, in itself a remarkable display of forbearance since orcas can be ferociously protective of their offspring. Apparently acting on his assessment of the problem, he swam under the stretcher and allowed one of the men to stand on his head. According to Tim, this too was remarkable since he had never been trained to carry people on his head. Then using the amazing power of his tail flukes to maintain a steady position, he provided the keeper a platform that allowed him to reach up and release the bridle so that the 420-pound baby could slide into the water and within reach of the keepers. Once the baby was back in the tank, Orky retreated to watch the rest of the procedure from the center of the tank.

Cooperation between captive animal and human can continue even when the animal has the option of escape. Of course, this has always been true of domestic animals, where such cooperation is to be expected. Dogs, cats, horses, and other domestic animals have been bred to be appealing, useful, and cooperative with humans. More remarkable are cooperative ventures between humans and animals who have no such domestic history.

One of the most astonishing examples of purely voluntary cooperation between animal and human in the wild comes from Santa Catarina, the southernmost state in Brazil. Karen Pryor visited the area in the late 1980s to see with her own eyes an entirely voluntary joint fishing effort by dolphins and humans that, according to town records, started in the middle of the nineteenth century. "I'd heard sketchy reports about the cooperative fishing," she says, "but when I went to Laguna and saw it for myself, my jaw just dropped."

In what she describes as a highly ritualized encounter passed on by generations of dolphins as well as generations of humans, the fishermen will line up in the shallow, murky waters in a bay near the town of Laguna. Up to ten dolphins will station themselves twenty feet or so farther out to sea. When the dolphins spot a school of mullet, they will dive and turn underwater and then reappear on the surface, swimming towards the fishermen. Just before they get within range of the nets, the dolphins will abruptly stop and create a surging surface wave that carries the mullet the last few feet towards the now braced fishermen, who cast their nets and haul in the panicked fish.

Pryor stresses that the abrupt stop and sideways roll to create a surge is not a part of any normal breathing routine, but seems to have been invented to achieve the effect of herding the fish towards the net. She also notes that there are no signals passed or other communication between the fishermen and the dolphins, and that the fishermen don't reward the dolphins with mullet. She hypothesizes that the dolphins take advantage of the confusion created by the nets to grab all the fish they need. Perhaps most extraordinary is that each net-casting session is initiated by the dolphins. The system yields a lot of fish, which wholly explains why both species have an interest in continuing this extraordinary example of cross-species cooperation. In one half-hour period, she witnessed six net-casting sessions as the dolphins drove mullet towards the waiting fishermen.

Not all the dolphins in the area support this endeavor. Disruptive dolphins will sometimes charge in aggressively and wreak havoc on the nets. (Perhaps they are jealous, or perhaps they know about the high mortality suffered by their fellow dolphins in the Eastern Pacific who serve as unwitting tuna spotters for fishing boats and consider the cooperating dolphins quislings.) When this happens, the fishermen say that the "good" dolphins will drive the "bad" dolphins off. (For a more complete account of this remarkable joint venture, see Karen Pryor's book *On Behavior*.)

Because there is sufficient motivation on both sides to keep the arrangement going in order to continue to reap the rewards, this long-term cooperation does not require intelligence on the part of either dolphin or human, only a desire to continue to receive rewards by repeating the behavior. But the humans are intelligent and aware that their ancestors stumbled on a good thing, and it is not much of a stretch to assume that to the degree that dolphins share awareness, they are also mindful that they are working with another sentient being.

The Laguna ritual is not the only example of cooperative fishing between humans and dolphins, nor even the only example of the two species fishing for mullet. Indeed, it is not even the only example of humans cooperating with a wild marine mammal. On the opposite side of the globe from Brazil, in southern Asia, fishermen have teamed up with wild otters, who also herd fish into waiting nets.

For her part, Karen Pryor is anything but a romantic about awareness in dolphins or most other animals. In recent years, she has tended more towards the behaviorist school, and she argues that many case studies offered by scientists as evidence for higher mental abilities can be explained by operant conditioning. She ranks dolphin intelligence below that of chimps, and above that of dogs. (The grass-is-greener theory extends to studies of animal intelligence: dolphin people tend to marvel at chimp intelligence, and vice versa.) On the other hand, even as she disputes some of the studies of animal language and intelligence, she argues that the surprising behavioral flexibility and ability to learn unearthed by operant-conditioning techniques in various animals suggests that intelligence flickers in a number of animals besides apes and whales.

In recent years, Pryor has championed "clicker" training, which uses the sound of a clicker and rewards to get animals to perform some desired behavior. The clicker serves as a marker, telling the animal that it has done something the trainer likes and that it will get a treat. While this sounds like a straight stimulus/reward technique, Pryor says that there is an "Aha!" moment when a bulb goes off in the animal's mind and it realizes that it is being rewarded for something it has done, and at that point the animal will start trying to think of what it just did.

For instance, she once worked with a show horse that had

a tendency to lay its ears back on its head during performances, giving the animal a beleaguered look no matter how
much it might have been enjoying the performance. Using
the clicker, she taught the trainer to make a sound and give
the horse a treat when it moved an ear. After a while, the
horse figured out that the clicker and the rewards had something to do with its ears, and it started moving them in various ways, first one up and one back, etc. The trainer only
responded when an ear was straight up. In just one session,
the trainer had taught the horse to keep its ears up. "When
we saw the horse experiment with its ears in different posi-
tions," says Pryor, "we were seeing it thinking."

Once the bulb has gone on, says Pryor, it stays on, and the
animal becomes engaged in thinking of ways it can cooperate with its trainer. At a logging camp in Washington State,
one clicker-trained workhorse named James spontaneously
began leaving the logs he dragged in neat rows after he observed that this seemed to be what the humans wanted. This
same horse tried to help his owner dig holes for fence posts,
helpfully tagging along with a shovel.

Sounding suspiciously like Donald Griffin for a profes-
sional behaviorist, Pryor believes that a lot more animals are
capable of having this "Aha!" experience than is commonly
credited. Using her clicker, she once taught her terrier,
Skookum, to do creative tricks, just the way she once taught
dolphins. For instance, she would click and reward him
when he first touched a chair, and then reward him for the
first thing he did after touching the chair. However, she
would then not click if he repeated that same motion, but instead click and reward him when he did something different. In this way, she guided the dog towards inventing his
own games, one of which, she said, involved a hilarious stunt
in which he got under a large cardboard box and moved it
around the room so that it looked for all the world as though
the box had some mysterious source of power.

Rob Shumaker agrees that operant conditioning can produce results but argues that many behaviorists miss the point of what is going on between the animal and the trainer. Where behaviorists believe they are conditioning the animal, what they are really doing is communicating. "A clicker or a target is a way of keeping an animal's attention focused on the conversation," says Shumaker, "but what is really important is the communication."

This conversation between animal and human is not always benign. Far more controversial examples of cooperation between humans and free-ranging animals have been efforts by the military to enlist marine mammals in surveillance and even warfare. Some of these efforts are ostensibly harmless. For instance, beginning in the 1980s, the Office of Naval Research sponsored research efforts to see whether sea lions carrying miniature video cameras strapped to their heads could be trained to follow and film whales. Dan Costa was one of the scientists involved in this effort at Long Marine Laboratory near Santa Cruz, California. The whole idea, he says, was to encourage the sea lion to do something it ordinarily does, and does many orders of times better than any human. Sea lions can dive to eight hundred feet, and their agile, familiar forms would not be perceived as a threat by the animals being trailed (although, if that animal was an orca, the sea lion would be perceived as dinner).

Using rewards, Costa and his team trained the sea lions to track an artificial whale in a tank. Then they began trying to transfer their training from a fiberglass form in a tank to the much larger real thing in the wild. Starting with simple videotaping, Costa hoped to expand the training so that the sea lions could be enlisted as research assistants—tagging whales and other sea life with telemetry devices, taking pictures, identifying the sex of whales (an exasperating and inefficient process for humans), photographing the sea bottom, and otherwise checking things out.

It is easy to see how all these activities might be turned to military uses, and indeed the Navy's use of marine mammals extends back decades to the Vietnam War. In the late 1980s, I explored the military use of dolphins in an article in *Time* magazine. The attempt to enlist dolphins in human wars has to be one of the most misbegotten examples of animal/human cooperation in the history of our species.

The Navy has long used dolphins to retrieve objects from the sea bottom and as scouts in underwater rescue missions. Inevitably, however, some bright light in the military decided that these highly intelligent, speedy animals with an extraordinary ability to map the underwater world could be turned into weapons of war.

Using its echolocation and acoustic imaging abilities, housed in the melon that gives the dolphin a nobly shaped forehead, and supplementing that with sensitive high-frequency hearing, the dolphin maps its world through noise. When it "sees" a human swimming, it essentially sees a skeleton. Its abilities are so highly developed that an orca (which is a very, very large dolphin) can stick its head into the mouth of a bay and, in a matter of seconds, send out enough clicks so that the animal knows whether there is anything in the bay that merits further investigation.

It is obvious why such abilities could get military planners drooling. Here was a highly flexible animal with underwater skills that are generations ahead of human technology and whose cost was an infinitesimal fraction of most underwater weapons systems. Even better, since dolphins aren't made of metal and do not emit a lot of heat, they would be more difficult to detect and counter.

In the late 1960s at the Naval Undersea Center in Point Mugu, California, and then later in Kaneohe Bay, Hawaii, the dolphins were taught through the use of rewards to attack objects with barbed darts delivered to the target by an apparatus attached to the dolphins' heads. The idea was to use dolphins as underwater sentries, protecting the American base at Cam Ranh Bay in Vietnam from incursions by enemy divers. The dolphins were to stick enemy divers with the darts, which were attached by sturdy thread to floats. Supposedly surface patrols would then snag the floats and reel in the divers.

It is unclear whether this grotesque scheme was ever put into action because the material relating to these missions is still classified. I did speak to some of the trainers. One of them said that a number of his colleagues quit when they finally realized what the animals were being prepared to do. One man who asked to remain anonymous still feels disgust at the mission. Said he, "The whole program was a hideous use of the most benevolent creatures I've ever had the chance to know. To the dolphins, it was all games."

The military mind being what it is, the idea of using dolphins for military purposes has repeatedly resurfaced in the years since the war. A number of dolphins were trained to protect the Trident submarine base in Bangor, Washington. In this case, the dolphins were not being trained to attack humans, merely to mark the location of intruders, but news of the program sparked protests and lawsuits by dolphin lovers. Among the many ironies of this program was that in the late 1980s and early 1990s, when it was launched, the most likely intruders to be detected were antinuclear activists and animal rights activists. There was and remains the possibility that a dolphin might help the Navy arrest a dolphin lover.

During the Iran-Iraq war, in 1987–88, five dolphins served as underwater guards protecting American ships. Again, during Operation Desert Storm in 1991, there were persistent rumors that the Navy was using dolphins as underwater sentries.

The United States was not the only nation that saw the opportunity to enlist dolphins in military activities. With typical brutality, the old Soviet Union reportedly killed over three hundred dolphins in the course of testing the animals as kamikaze agents to carry explosive charges during exercises in the Black Sea, according to Doug Cartlidge, a former dolphin trainer now affiliated with a British organization called the Whale and Dolphin Conservation Society. As reported by the English newspaper *The Independent* in 1998, Cartlidge visited the formerly secret Soviet facility in Sevastopol, where he saw models of dolphin parachute harnesses (one method the Soviets had of infiltrating their finny agents; another was dropping them out of helicopters from a height of fifty feet). Cartlidge also claims that the Russians trained dolphins as killer agents, and that they got around the animal's natural reluctance to cause distress by devising a titanium clamp that could be attached to the dolphin's snout. If the dolphin simply nudged an enemy diver, the clamp would then affix itself to the diver's body, and once it was affixed, a second device in the apparatus would inject the diver with a fatal dose of highly compressed CO_2 gas. With the collapse of the Soviet Union, funding thankfully disappeared for this macabre program. Alas, according to Cartlidge, some of the Russian Navy's cadre of marine mammal specialists have gone into the business of capturing wild Black Sea dolphins for sale to the aquarium trade.

Dolphins can certainly be aggressive with their own kind, and they are not shy about attacking threatening predators, but it is unclear whether they would knowingly kill a diver. There is no question that a dolphin is smart enough to distinguish between a dummy in a pool and a living being. Moreover, an animal that can tell whether a swimming human is pregnant by giving her a sonogram will also be aware of when it is inflicting injury and pain on another creature. Yet Karen Pryor says that it is entirely possible to train a dolphin to attack strange humans. While they might see the trainer as part of their group, she says, whatever human/dolphin bond they developed would not extend to other people, particularly if they were being rewarded for aggressive behavior. Until reports of their activities in warfare are declassified, we can only hope, for their sake, that they proved to be rotten soldiers.

More recently, the U.S. Navy has gone public with its latest version of the marine mammal program. Both the Navy and the Marines have trained a number of dolphins for minesweeping operations. Because most underwater mines are detonated through contact with metal, the dolphins can approach them without risk. When they detect a mine, they have been trained to release a buoy that floats to the surface, alerting Navy divers. The Marines use the dolphins in very shallow water, where their maneuverability gives them a huge advantage over man-made vessels. In 1998 the Navy unveiled the dolphin minesweepers in an international naval exercise called RIMPAC off the coast of Hawaii, and then in the Baltic as part of an exercise to help the nations of Latvia, Lithuania, and Estonia rid themselves of an estimated thirty thousand underwater mines remaining from the Cold War era. The Mark 7 Marine Mammal System, the ridiculous name given to the dolphin minesweeper unit, failed to find a mine during the four-day exercise, and as reported by the *Washington Post*, one of the units went AWOL for three days. Like so many other lonely soldiers stationed far from home, he was looking for a girlfriend.

The U.S. Navy takes great pains to assert now that dolphins are not used in offensive operations, and that it treats the animals in an exemplary fashion. But by pursuing these military programs, the United States is validating a type of marine mammal arms race that inevitably leads to the barbaric and obscene misuse of these animals in programs such as those allegedly carried out by the United States in Cam Ranh Bay and the Soviets more recently. Given the dolphin's intelligence and awareness and its ancient reputation for saving shipwrecked seamen (even as I was writing this chapter, there were news reports of a group of sailors who, forced to abandon their ship and tread water while waiting to be rescued, claimed that they were protected from sharks by circling dolphins), it is nothing short of evil to enlist these animals in what dolphin researcher Stephen Leatherwood called our most "vile and reprehensible activity": warfare.

Moreover, those clever minds who decided to capitalize on the dolphin's underwater virtuosity have created a cloud of suspicion that hangs over any strange dolphins in the vicinity of a military base or theater of conflict. If these programs continue, some paranoid military types are going to start killing stray dolphins that wander near sensitive areas on the grounds that it is better to be safe than sorry.

Humans have used other animals in conflict as well, of course. The Romans, the Indians, and the Thais, among others, have trained elephants for warfare, and their role extended beyond carrying soldiers to crushing the enemy. In fact, one form of execution once used in India was to have an elephant step on the head of the condemned. Again, Karen Pryor expresses no surprise that a smart animal like an elephant can be taught to kill humans, noting that it is all a matter of training.

I suppose it is overly sentimental for us to expect that a wild animal should be more respectful of human life than we are. The vast bulk of cooperation between animal and human in situations of conflict involves domestic animals. Guard dogs, drug-sniffing dogs, and police dogs are so integrated into the police and security infrastructure of humanity that their appearance passes without comment, except on those occasions of notable heroism. If man's best friend can be turned into a killer, we should not be surprised that an elephant or a dolphin can too.

Still, it would be both unfair and wrong to suggest that most human/animal cooperation has to do with conflict and law and order. Humans have long turned to animals for help in therapy and healing, and these interactions also often suggest awareness and even empathy that extends across the species barrier (I will have more to say about empathy in a later chapter).

Much of the use of animals in therapy is predicated on the observation that people who have undergone severe injuries or trauma sometimes respond more to animals than people. Harley de Swine, a Vietnamese potbellied pig, is a good case in point. He "worked" in a ward for patients with head injuries at Hemet Valley Medical Center due west of Palm Springs, California. Therapists would determine which side of a patient they wanted to stimulate and then bring in Harley, whose job was to gently get the patients to pay attention. Harley could often get a response from head-injured patients where people failed, but what also really impressed the therapists was the enthusiasm the pig brought to his job. Darrian Lundy, Harley's handler, says that he would run to work in the morning, greet the staff, and drag Lundy along toward the ward. At the end of the day, he had to be prodded to leave.

Cats, who have carefully and successfully maintained their aloofness despite thousands of years of breeding, also on occasion have displayed great tolerance and sensitivity. Maureen Frederickson of the Delta Society counseled an abused woman in the Seattle area whose cat would present her with two mice every night. (I know, I know—lots of cats do this, including mine. The experts say that cats are trying to teach us to hunt, but I continue to believe that such gifts are their way of squaring accounts by contributing to the family larder. I also fear that we continually disappoint our pets when we fail to eat the mice and birds.) When the woman finally mustered the will to flee her abusive husband and moved to the city, the cat could no longer find mice. So it started bringing her two pinecones every night. Maureen wonders whether the cat was offering the gifts as a form of solace to ease the woman's obvious suffering.

Cats are also frequently used to brighten the days of those confined to nursing homes. Maureen notes the story of one big tomcat show remarkable tolerance for an aged senior at a home in Ontario, Canada, a man whose body was so twisted by arthritis and tension that he could not open his hand more than to make a clenched U. The man, who had been institutionalized because he was unable to function following unspecified traumas during World War II, at first simply rested his clenched hand on the cat. Mona Sams, an occupational therapist then working at the home, taught him to pet the cat. Sams says that the patient began to open up through interactions with the cat. He grew devoted to Yoda, as the cat was called, and, says Sams, within three weeks he would reach out his arms when he saw the animal. Yoda, named for the wise Jedi elder of the *Star Wars* saga, offered the staff a way to reach a patient who for previous decades had been almost completely withdrawn.

Sometimes animals will show tolerance for the afflicted to the point of overruling their natural instincts. In Belgium, the horses in a program of therapeutic riding for autistic kids tolerate behavior from these children, such as sudden fits of violence, that would have them kicking, bucking, and fleeing in ordinary circumstances. The horses will let the children walk under their bellies, which would ordinarily spook the animal, as anyone who has worked with horses knows.

If most of these examples of cooperation involve humans taking advantage of animal skills and sensitivity, there are at least some cases of a wild animal showing some initiative in adapting human skills for its own purposes. Roughly a decade ago, some Florida scientists who were using food pellets to attract fish at a particular spot in a bay. An ecologist told me the story of how a local heron observing the scientists at work figured out that if it hovered nearby, it could catch the fish when they came up for the pellets. After some days of this, either the fish got wise, or they had all been eaten, because the scientists would drop pellets and nothing would come to the surface. Taking matters into its own beak, the heron picked up a pellet, moved about fifty feet down the shore, dropped the pellet in the water, and when a fish surfaced, grabbed and ate it.

Finally, there is the issue of cooperation between captive animals. I opened this book with the story of two elephants at the Bronx Zoo who figured out that by trusting each other and taking turns eating treats they could prolong their time outdoors. In a later chapter, I will offer some examples of chimps and gorillas cooperating in the execution of escapes, but apes can also cooperate with one another in other kinds of efforts. The renowned primate expert Robert Yerkes documented an experiment in which two chimpanzees had to pull equally hard on two handles to get access to a food reward. One of the chimps was highly motivated, but the other seemed indifferent to the task. In a moment caught on film, the go-getter chimp grabbed the other chimp and turned him around, as though to say, "Get back to work."

Natural selection crafts a species so that individuals can make it on their own or by fulfilling their role in a group. I know it is anthropomorphizing, but it is perhaps arguable that predators and all manner of animals feel some sense of satisfaction or pride when they do well what nature has equipped them to do. As noted earlier, there must be some emotional reward when a male chimp or elk or lion successfully challenges another for dominance, and it is probably not that different from the rush felt by any successful human athlete or warrior. After all, that rush is a product of our biological makeup, and physically we are not so very different from the other members of our primate extended family, which includes gibbons, bonobos, chimps, gorillas, and orangutans. (Actually, maybe we are a little different from gibbons—the one truly monogamous ape.)

For most mammals, the main challenges of life are securing a livelihood, competing within the group, and finding a desirable mate. At least for males, success at this third ambition rests to a great degree on meeting the first two challenges. Being good at any of these tests probably makes an animal feel complete and good. Unfortunately, even the most sensitively designed captive situations cannot help disrupting these occupations and preoccupations. Which is to say that for better or worse, captivity takes away most of an elephant's opportunities to feel good about being an elephant. It largely eliminates the challenges that make snow leopards take pride in their power, grace, and skill, and it deprives an orca of an oceanscape big enough to fulfill orca dreams. Finding some meaningful way to relate to their strange new world may partially compensate for the loss of freedom that captivity entails.

ORANGUTAN ENGINEERS AND NUT-CRACKING CHIMPS

Tools and Intelligence

ALWAYS use the right tool for the job, the saying goes. Nina, a female gorilla at the Woodland Park Zoo in Seattle, took this concept to a new level when she used her own baby as a tool to get food a couple of years ago. Female captive gorillas can be responsible and protective mothers, but their nurturing skills can be tested when their babies are given an appetizing treat. Such was the case with Nina and other gorillas at the zoo who regularly appropriated for themselves treats meant for the infants. Taking pity on the babies, the keepers built a separate enclosure inside the holding area with an opening that only a very young gorilla could fit through. After they distributed the treats to the adults in the colony, the keepers would then put some fruits, cooked yams, and carrots in the babies' area. Never one to give up a cooked yam without a fight, Nina figured out the expedient of holding her baby by the ankles, letting her crawl through the small aperture until she reached the food, and then pulling the baby out and snatching the treats before the baby could eat them. Nina gets high marks for ingenuity if not mothering skills.

It is still unclear where tool use and toolmaking fit in the pantheon of higher mental abilities. In the science fiction classic *A Mote in God's Eye* by Larry Niven and Jerry Pournelle, voyagers from earth reach a world in which the dominant species has subdivided and evolved into several highly specialized subspecies, each optimized to meet the needs of a society whose dominant trait is technological genius. There are negotiators, diplomats, drivers, engineers, warriors, and then there is a class of small animals, dubbed "watchmakers" by the visiting earthlings, known for their propensity to fix and improve any appliance they are handed. These deceptively appealing creatures do not seem to have a highly developed intellect or language, but they are frenetic, unstoppable builders and fixers. Their intelligence is expressed through their hands.

Apart from the main narrative, the book provokes some interesting thoughts about the nature of technology. Is *Homo faber*—man the builder—the antecedent of *Homo sapiens*—man the thinker—or did these abilities arise together? Are there antecedents and analogies for human material culture in other animals? And as in the fictional case of the watchmakers, can a creature possess great technological gifts without displaying language and the other higher mental abilities of the planet's most adept tool maker and user?

Apes seem to grasp how to combine the objects in their environment to achieve some goal. As long ago as the 1930s, Wolfgang Köhler, one of the pioneers in the study of animal behavior, demonstrated that apes could figure out how to stack two boxes to make a platform high enough to reach something out of grasp when standing on just one box. Zookeepers around the world have watched orangutans and other great apes stack barrels, make bridges out of logs, and use all sorts of implements to effect escapes. At the Woodland Park Zoo, keepers had erected encircling hot wires to protect trees that were being destroyed in the outdoor gorilla exhibit. Jumoke, one of the adult females, would put logs and tree branches over the hot wire so that she could get at the tree.

When Jumoke had to be separated from Alafia, the father of her baby, she was placed in an adjoining enclosure bounded by a large, smooth-walled moat. One volunteer told me that she saw Jumoke prop herself up with the help of another gorilla and hold her baby so that its dad could see the little gorilla from the other enclosure. Unfortunately, I could not confirm this incident with the keepers because they were not present at the time.

Apart from the baby's father, most of Jumoke's family group was in the adjoining area (the old gorilla exhibit area, which was being replaced with the new enclosure that now housed Jumoke), and she was desperate to get back. Wolfgang Köhler would have been proud of her improvisations during her various attempts to rejoin her group. She tore apart climbing structures in her effort to get large boards that she could prop against the moat wall. Foiled on this attempt, she ripped limbs off trees towards the same end. Standing on the roof of the ape exhibit, Violet Sunde interrupted one of these attempted escapes. Noticing that Jumoke was carrying a big limb towards the moat, Violet said, "Jumoke, what are you doing?" The gorilla instantly dropped the limb. Keepers cannot keep a constant watch, however, and Jumoke did succeed twice in getting out. On one of these occasions, the keepers found her at the door of the old exhibit, confirming that the gorilla's intent was not so much to escape as to get back to her family and friends.

Scientists exploring these issues stress that success at using tools does not necessarily imply insight into their purpose. By merely blindly imitating the actions of others—using the simple rule "do as he does"—an animal can achieve a certain measure of success in using tools. In Bujumbura, Burundi, Geoff Creswell, who was then helping Jane Goodall in a project to better the conditions for orphaned and confiscated chimps, told me of one amusing instance of serial imitation.

When I met him in 1991, Creswell was taking care of a little female chimp named Ali. As with any infant, he tried to deal with her tantrums by distracting her. During one such tantrum, he started digging a hole in the sandbox behind his house. Sure enough, Ali came over and peered at what he was doing. Then she pulled his hand out and peered into the hole. Then she sniffed his fingers, but he just pulled away and kept scooping sand. After a few more seconds, she started digging her own hole, right next to his. Then, a minute later, the dog came over. After observing the human and chimp digging for a minute or so, the dog decided to join in, and began digging his own hole.

As far back as 1925, Wolfgang Köhler argued that understanding requires something more than exact imitation, namely, a grasp of what another's action means. Today, this is referred to as "program-level imitation," meaning that, as Köhler described it, child, adult, or animal understands the purpose of a series of actions, rather than looking at the actions as a type of magic ritual that, when performed exactly, yields a given result. In Hawaii, Louis Herman points out that the dolphins Akeakemai and Phoenix watched humans throwing a Frisbee and then learned to throw it with their mouths. Not having hands, this was their only choice, but it suggests that the dolphins understood that by imitating the motion of a sweeping throwing gesture, they might achieve the same result as a human throwing a Frisbee. Chantek's use of his lips to tie knots is, as noted in an earlier chapter, another example of program-level imitation.

There is also no question that apes and other "smart" animals also resort to straight mimicry when they cannot figure out how something works, even if they understand its purpose. Helen Shewman remembers one memorable escape in which the orangutan colony got up to the roof of their old enclosure. The keepers had brought out fire hoses, not to blast the red apes but to dissuade them from descending from the roof in a way that would take them down into the lions' or grizzly bears' areas that adjoined their complex at that time. Towan, the dominant male, took offense at this, however, and, spotting a fire hose on the roof, unspooled it and, outnumbered and outgunned but still unbowed, pointed it back at the keepers like Fess Parker in the Disney version of the fall of the Alamo. Fortunately for the keepers, Towan did not have the wrench needed to turn the water on. If he had, he probably would have figured out how to use it.

Towan provided another comical example of this near-miss imitation when he tried to figure out Helen Shewman's two-way radio. Given enough time, it is entirely possible that the patient orangutan would have figured out how to work the device. Talking on the radio, however, is another matter. In the case of the orangutan, it would have required either a few million years of evolution or the invention of a speech synthesizer.

Some of Towan's toolmaking and tool use serves no other purpose than to amuse himself and annoy his keepers. Two feet above the wire-mesh cage top in the indoor orangutan exhibit there used to be a set of fluorescent lights secured by butterfly nuts. Towan broke off a steel rod in his own cage, and using this probe, he reached through the wire mesh and unscrewed the butterfly nuts securing the light fixture. When Libby Lawson, one of the keepers, came on the scene, a small crowd had gathered to watch the orangutan bash the lights now sitting on top of the mesh. Using her most severe voice, she yelled "Stop," and Towan instantly ceased his depredations. She went to get reinforcements, and when she returned, Towan was standing equably in his cage with the electric cord for the light slung over his shoulder. Needless to say, electricians moved the light fixture higher and got rid of the butterfly nuts.

One of the mysteries surrounding orangutans is why these animals, consummate tool users in captivity, show very little evidence of tool use or material culture in the wild. Carl van Schaik of Duke University has documented that orangutans in Borneo fish for termites, although they use a different approach than the chimps at the Gombe Stream Reserve. In

Borneo, they put small sticks in their lips and insert the sticks into holes in trees to draw out termites.

It's possible that orangutans may resort to tool use in the wild more frequently than has yet been documented. In captivity, they will use vines as lassos to snag or reach plants and bring them within range of their arms. It is such a natural action that it is entirely possible they occasionally do this in the wild as well.

In captivity, orangutans approach tool use in very different ways than other animals, or even other apes. Rob Shumaker has noticed that orangutans, particularly the males, tend to use their lips where other great apes would use their hands. In using their lips, they are simply taking advantage of muscles adapted for fine motor tasks. They are far more adroit than most humans at controlling their lips and tongue. Says Shumaker, "You know that old bar game of putting a cherry stem in your mouth and tying a knot? That's nothing for an orang." Shumaker also says that an orang will take an apple in its mouth, move it around, and then spit out the entire skin.

On the other hand, an adult male orangutan's hands are simply too big for many tasks designed for human-sized fingers. Humans are bipedal, and our big muscles are in our legs. Orangutans spend much of their time in trees, and, notes Shumaker, the burden of locomotion falls on an orangutan's arms, which are astonishingly big and strong. Apart from being too big for some of the fine tasks required in tool use and toolmaking, an orangutan's thumb and forefinger are separated more than a chimpanzee's, further reducing the utility of its hand for fine motor tasks.

An example of orangutan insight into the nature of tools and machines came out of Lyn Miles's language experiment with Chantek. On one occasion, Lyn wanted to give Chantek an orange drink, but the can would not fit through the bars. A frustrated Chantek finally pointed to a drinking fountain in his cage and signed, "Put there." Lyn thinks that he was asking her to put the drink into whatever supplied the water in the fountain so that it would come out in his cage. If so, the example shows fantastic ingenuity, if some naïveté about the nature of plumbing.

At the Tulsa Zoo, one of the female elephants used her trunk and some understanding of hydraulics to get an apple that was out of reach. The apple was in a pool in the elephant enclosure, but inaccessible because a crust of ice had formed on the water. The female elephant put her trunk through a hole in the ice and simply blew air at the apple, moving it to a point where she could retrieve it.

To understand the purpose of an action rather than simply mimic a series of movements, an animal needs some form of internal representation of what is going on. Certainly an animal can test some types of propositions in its mind without having the capability of generating and understanding symbols. Donald Griffin pointed out in Colorado that an animal might visualize the outcomes of various actions while planning what to do when it sees a poisonous snake. When, however, the actions are displaced in time and space from a goal, the animal must develop a plan, which in turn requires that the animal must develop a program for the series of movements that will achieve its goal.

Take, for instance, one astonishing experiment I witnessed some years back with the pygmy chimp Kanzi. Following in the work of Nicholas Toth, an Indiana-based paleontologist, Rose Sevcik conducted a study of Kanzi's use of stone tools in which the bonobos had to find a flint stone, make a knife edge, then use the knife to cut a string, open a box, and get a key. Only then could Kanzi use the key to unlock another box to get a treat.

Sue Savage-Rumbaugh showed me how the experiment unfolded: Kanzi watched as a pomegranate was placed inside a box. The box was then locked, and the key to the lock was placed in another box that was tied shut with a sturdy string. The test began when Savage-Rumbaugh told Kanzi that rocks were outside in the playground. To communicate with the ape, Sue used a handheld keyboard that generates words depending on which symbol is pushed. Using the same keyboard, Kanzi punched a symbol, and a computer-generated voice said "key." Kanzi was asking for a key to open the gate. Striking to me were his concentration and his amazingly quick eyes as he alternately looked at Sue and the box holding the pomegranate.

Outside he picked up a rock, then quickly returned to the room where the box was placed. He hurled the rock against the floor, breaking off chips of flint. Picking up a sharp piece, he immediately set to work cutting the string. Once he had the key, he opened the lock and retrieved the pomegranate.

In the several tests I saw, Kanzi also retrieved a ball and grapes cached in the same way, and only once screwed up, trying the key on the wrong lock. The first time he saw Kanzi do this, Nick Toth said, "For a Stone Age archaeologist like myself, seeing this was almost a religious experience." He also remarked that for the first time he could compare the abilities and tool use of a modern ape with those of 5-million year-old hominids. In fact, Indiana University was sufficiently impressed with Kanzi's accomplishments that they gave the pygmy chimp an award for contributing the most insight into the origins of technology.

There have been numerous other experiments exploring the degree to which apes and other animals understand the utility of tools. As mentioned in an earlier chapter, at Think Tank, one of Rob Shumaker's experiments involved a situation in which an orangutan had to give Rob an appropriate tool if he was to fetch cookies or other treats that the ape wanted. He would ask the orang for help, but not for a specific tool. On one occasion, Indah, a female orangutan, noticed that the appropriate tool was in Rob's cage, and also that he could not see it. She reached through the bars, grabbed the tool, and handed it to him, demonstrating that she understood both that he was unaware that the tool was available to him and that he needed a tool if he was to fetch cookies for her. This and the other examples cited earlier suggest to Rob that the orangutans were using insight rather than blind trial and error in their use of tools. "They will solve exactly the same problem [how to get Rob to get them a treat] in a novel way," he says. They seem to realize that a tool could help them achieve their ends, but more than that, they understood that a particular task requires a particular-length tool and adjusted their behavior accordingly.

Animals make and use tools in the wild as well, and you don't have to trek to the wilds of Borneo to see it. Herring gulls will grab a shellfish, fly above a rock or pier, then drop the mollusk to break its shell. Any number of animals build structures with beaks, claws, hands, and pincers. Bowerbirds build bird mansions; termites build mounds that are architectural wonders in their ability to maintain stable temperatures in their interior despite the extreme swings of temperature in the African veldt. There are even rain forest ants that practice agriculture. Leaf-cutter ants lug bits of leaves into the interior of their mounds, where they use them as compost to grow mushrooms (or fungus if we must be literal). These ants also have a livestock operation of sorts, shepherding aphids for their nectar. Even if such behavior is genetically programmed, it is an ingenious evolutionary solution to the problem of ensuring a stable food supply.

While there is little need to invoke consciousness to explain ant architecture and agriculture, awareness becomes more likely—if not necessary—in the material culture of some of the higher mammals. In the Tai Forest in the Ivory Coast, Christophe Boesch, who has been studying the area for more than twenty years, led me to the base of a buttress-rooted tree, where I saw what looked for all the world like a primitive food-processing center used by a hunting-and-gathering people. Lying around were well-worn wooden implements and dozens of nutshells. This was indeed a workspace, but the workers were chimps who set panda nuts and other shelled foods in hollow knotholes on tree roots before breaking the shells with stone or wood hammers.

During the nut-cracking season, the forest resounds with the sounds of hammering. It sounds so much like construction that one visitor asked Boesch who the carpenters were. During one of my forays into the forest, I tried to use the chimp-made tools to crack a nut. Using a two-handed grip on a rock (the grip favored by chimps), I first made the mistake of placing the nut in too deep a hole, and then, once I found the right-sized notch for the nut I had in hand, I mashed the kernel by hitting it too hard. Chimps would not make my mistakes, Boesch assured me, and then told me of watching a female chimp who seemed to instruct her son in the art of nut-cracking. She would place a nut on an "anvil" (as Boesch referred to the notch where the nut is placed for cracking) and then leave a hammer beside it for the young chimp to use. The mother, whom Boesch called Salome, once interrupted her son as he was about to strike an improperly positioned nut, swept the anvil with her hand, then replaced the nut, now properly positioned, for her offspring to hit.

The Tai Forest is one of the last remaining patches of relatively pristine rain forest in the Ivory Coast. Despite its special status, it is threatened on all sides. When I was there, the upheavals in neighboring Liberia had driven refugees into the towns adjoining the forest, some of whom tried to poach game, tempted by the relatively plentiful supplies of bush meat in the forest. More recently, an outbreak of the deadly Ebola virus swept through the chimp population, wiping out a number of animals that Boesch had studied and come to know since he first began studying the groups in 1979.

The forest is dank and dark, and I spent most of my time sweltering in the heavy dark woolen clothes that Christophe insists visitors wear when following him on his rounds (in part to help the chimps tell the difference between "safe" hu-

mans and poachers). The dominant smell is of rotting vegetation, but every now and then we would walk through a
pocket of fragrant pollen hanging in the dripping air. One of
the striking things about the Tai Forest is how few stones
there are. According to Boesch, this scarcity puts pressure
on the chimps to remember where a handy stone might be
when it is required, for instance, to crack the tough-shelled
panda nut. Christophe says that the chimps seem to know
the most direct route to the nearest stone. To Boesch this implies that the chimps maintain a mental map of the area,
something he claims does not occur in human children before age nine.

Nor are the Tai chimps unique. Nut-cracking takes place
throughout West Africa. It was first documented by Dr. Yuki-
maru Sugiyama of Kyoto University on the slopes of Mount
Nimba in Guinea in 1976 when he watched an isolated
colony of chimps use stones to crack oil palm nuts. In neighboring Guinea, Tetsuro Matsuzawa watched a female chimp
construct a stable stone platform on which she could crack
nuts. At first the nuts slid off the tilted stone before she could
crack them, and so she shoved a smaller stone under one
edge of the larger stone so that the top surface was level. In
this same forest, researchers have documented male chimps
shoving palm fronds into the tops of oil palm trees to get at
the pith. As far back as the 1960s, Jane Goodall broke the
news that chimps fashion sticks to fish for termites. What
this all means is another issue entirely.

Even stripping leaves and sprouts to make a simple fishing stick for termites requires some image of the desired object as the chimp is fashioning the tool. It also requires some
control over the hands and the ability to program the movements those hands make.

In this sense, organizing a series of actions the way wild
chimps do during nut-cracking and the way Kanzi did to get
at the pomegranate can seen as the physical equivalent of

forming a sentence in human language. The order of the motions was important just as word order is important in a sentence, and the actions were abstract in the sense that the end result of getting at the pomegranate was not implicit in actions like breaking a stone to create a knife, that were necessary preconditions of getting at the fruit. In other words, something akin to a syntax is needed to organize a complex task. The actions involved in stabilizing a platform, for instance, take priority over other movements involved in nut-cracking.

In fact, the parallels between tool use and the structures of language are even replicated at the neurological level in the human brain. In humans, the word *aphasia* describes various impairments of language abilities. Usually, these disabilities result from little strokes or brain damage of some sort to the centers that control language. Certain forms of aphasia leave intact a patient's ability to say words, but destroy his or her ability to meaningfully put words together in sentences. *Apraxia* is the word used to describe impairments of a person's ability to control physical movements. It bears noting that just as some aphasias allow people to say words but not link them together in sentences, some forms of apraxia allow people to make individual actions but not link them together into a useful pattern. In some cases, the same injury to the brain destroys both the ability to link words and the ability to link actions.

Twenty years ago some scientists interpreted this as evidence that language and technology had a common origin in the ability to program actions. I gave some attention to this idea in my book *Apes, Men, and Language,* but interest in the thesis has languished until just recently. In the March/April 1999 issue of *American Scientist*, psychologist Michael Corballis of the University of Auckland in New Zealand revives this hypothesis, first proposed in the seventeenth century by the French philosopher Etienne Bonnot de Condillac.

Apart from evidence cited twenty-five years earlier by anthropologist Gordon Hewes—young children tend to point and gesture before they speak, the spontaneous emergence of sign languages among various deaf communities, etc.—Corballis cites more recent work imaging brain activity with magnetic resonance equipment. Parts of a region of the brain called Broca's area are activated in similar ways when someone speaks and when they make gestures. Corballis speculates that hand gestures began to take on meaning when hominids became bipedal roughly 5 million years ago, and that about 2 million years ago human ancestors used both gestures and vocalizations to communicate with one another.

The question of when spoken language reached its present stage of development is as contentious as any other matter dealing with human prehistory. Some scientists argue only *Homo sapiens* had both the physiological and neurological capabilities necessary to control lips and tongue and generate the range of sounds used in modern spoken language. Moreover, the gestural-origins hypothesis speaks mostly to the question of how speech evolved, and less so to the evolutionary pressures that caused this trait to flower. It is possible that in searching for a way to communicate as hominid societies became larger and as food gathering more complex, early man took advantage of a previously developed ability to form propositions through hand movements.

Orangutans throw a challenge at all these theories. Males tend to be solitary in the wild, so they have far less need of the politics that animals with highly developed mind-reading and communicative skills favor. Shumaker says that most orangutan interactions are one-on-one. They do not engage in cooperative hunting or group warfare, which would also select for the development of higher mental abilities in some mammals. They use tools far less than chimps do. Yet in captivity, they prove to be better techies than any other animal, and they also hold their own with respect to any other measure of intelligence. As Rob Shumaker says, "If you force them to perform tasks adapted to a human or chimp physique, they might not do as well, but if you broaden the standard to how they perform on equivalent tasks, they do as well or better than the other apes."

The gestural-origins hypothesis may not answer the mystery of why orangs have proved so good at human tasks in captivity (maybe orangs take special delight in frustrating cognitive scientists, just as they enjoy outwitting keepers), but it does help us understand how higher mental abilities for one purpose may prove useful in other applications as well. Free-living orangutans inhabit a wildly diverse ecosystem where they have to learn the placement and fruiting patterns of a bewildering variety of trees. The biologist Peter Ashton once documented five hundred different species of tree in a single hectare of rain forest on Borneo. A wild orang also has the chore of monitoring the movements of a widely dispersed group of females and the activities of competing males. Sophisticated cognitive abilities produced by one set of evolutionary pressures may prove useful in other applications as well.

Dolphins would also seem to challenge the gestural-origins thesis. Lou Herman's work demonstrates that they are pretty good at decoding complex human sentences, but lacking hands, they could not program gestures even if they wanted to.

Indeed, the dolphin raises all sorts of questions about the origins and nature of intelligence. It has a very large brain, which by many of the measures we use to assert our own intellectual supremacy—brain-to-body-weight ratios, folds in the neocortex, etc.—should make it a very smart cookie. According to Peter Magone, a marine biologist, one of the more interesting aspects of the dolphin brain is that its sophisticated cortical development evolved on top of primary cortical structures that have more in common with the brains of dim-witted insectivores like armadillos, anteaters, shrews, and sloths than they do with the brains of primates, the smartest of the terrestrial animals.

This merits a brief digression into a phenomenon called convergent evolution. Convergent evolution is the phrase used to describe the tendency of evolution to converge on a common solution to problems like flight in very different creatures. Some of the oft-cited examples are the toucans of South America and the hornbills of Africa and Asia, which, though they evolved many thousands of miles apart, developed very similar large beaks suited for eating the fruits and seeds of the rain forest. William Conway, president of the Wildlife Conservation Society, notes that the auk, a now extinct bird of the extreme north, and penguins, confined entirely to the Southern Hemisphere, developed similar physical structures in response to the selective pressures of frigid marine climates. It is also possible that evolution converges in the unfolding of higher mental abilities. If it turns out that similar physical characteristics in dolphin and primate brains produce similar higher mental abilities, perhaps there is an optimum structure for intangible abilities like awareness, just as nature seems to converge on similar designs when adaptive pressures push an animal towards flight or burrowing.

Is it possible that despite the dolphins' 40-million-year separation from life on land, nature could create remarkably similar higher mental abilities in creatures evolving in utterly different environments? If so, how many other intelligent creatures have come and gone, or are out there? If nature can create intelligence once (and it did so in hominids in just a few million years), perhaps it has done so many times in many creatures. The 200,000 years or so that human intelligence has existed is small enough to be a rounding error in the long history of life on the planet, and intelligence is an attribute that may equip a species for a meteoric rise and fall rather than long-term survival. Who knows? We like to think of evolution as progressive, but it is entirely possible that creatures with abilities as advanced as ours have come and gone before us. And it is possible that various animals developed similar intellectual abilities through entirely different sets of circumstances.

ESCAPE FROM TOPEKA...

And Omaha, and
Brownsville, and...

ONE activity assiduously pursued by animals in captivity exhibits many of the various expressions of intelligence that scientists spend their careers trying to elicit or observe in controlled studies and in the wild. In pursuit of this hobby, animals have demonstrated deception, toolmaking, program-level imitation, and mind reading. But instead of encouraging these pursuits, here is a case where the humans involved find such expressions of higher mental abilities most unwelcome, and in fact do their best to prevent the activity in question. I'm talking, of course, about escape attempts, a favorite pastime of a number of zoo animals—but a singular obsession of orangutans.

The Wallenda family had a tradition of high-wire walking. New York's Grucci family has staged fireworks exhibits for generations; Britain's Huxleys have produced prominent scientists and thinkers for two centuries. And then there is Jonathan, an orangutan now at the Cleveland Metropark Zoo and a master escape artist, and his progeny, who have been escaping from zoos across the country. Rudi, the orangutan's daughter, followed in her father's footsteps right out of exhibits. She spent some time at Lowry Park in Tampa, Florida, but staged so many escapes and attempts that she was sent back to Topeka, where she had been raised.

Even as I was in the midst of a series of conversations with one of Jonathan's former keepers in December 1998, he told me that two days earlier Jonathan's son Joseph had staged an escape from the Kansas City Zoo in Kansas City, Missouri. This escape was simplicity itself. Joseph had simply taken a rubber tire he had been given to play with, laid the tire across the electrified "hot wire" that encircled the exhibit, and climbed out. Then, when I was revising this chapter, word came that Jonathan himself had again escaped, this time from Cleveland's Metropark Zoo.

Jonathan adapts his escape strategies to the situation at hand. Some years ago, when Jonathan was at the Los Angeles Zoo, he was let into a new enclosure designed to give the orangutans more free space. One of the features of the new exhibit was a nice tree in the center of the enclosed area. On one of the first occasions he was let into his new home, Jonathan surveyed the surrounding walls, looked at the tree and then in front of the onlooking keepers, dignitaries, and passersby, ripped the tree out of the ground, placed it against the wall, and climbed out.

After Los Angeles, before he came to the Topeka Zoo in Kansas, Jonathan spent some time in Buffalo, New York. There he was held in what one keeper described as an "archaic" exhibit which had a glass-fronted wall for public viewing, and bars in the back that separated him from the service area. Barring negligence, escape was out of the question, and so Jonathan amused himself through other forms of mischief. Given burlap bags to play with, he would whip them through the bars and snag steam pipes that ran along the wall of the service area. Once he had snagged a pipe, he would pull it off the wall.

A cognitive scientist could have used Jonathan's attempts to *épater* his keepers as a one-orangutan study of animal tool use and tool invention. For instance, when the keepers wisely decided to take away the burlap bags and give him a cardboard box to play with, Jonathan didn't miss a step. He tore apart the boxes, fashioned the cardboard into spikes, and then used the spikes to break the fluorescent lights set in the ceiling of the cage area.

In fact, by the accounts of those keepers who dealt with him, Jonathan has displayed almost every behavior discussed so far in this book. He is an inveterate trader (showing a partiality to unfiltered cigarettes when he gets his hands on something his keepers want). If a simple strategy like pulling up a tree cannot get him out of an exhibit, Jonathan calls on more sophisticated abilities. Indeed, his most memorable escape showed planning and patience worthy of Tim Robbins, who in the role of an unjustly accused murderer spent years planning his prison breakout in the movie *The Shawshank Redemption*. To understand the many moving parts of Jonathan's most artful escape attempt, it is necessary to set the scene.

The "Discovering Apes" building at the Topeka Zoo consists of an outdoor exhibit area connected by a hallway to an inner holding area. This area consists of a series of cages arrayed in radial fashion around a central area. Off the hallway connecting the exhibit and the holding area is a service area that contains a small cage for holding apes, and this area connects by a doorway to stairs leading to a lower level. What are called "guillotine" doors control access to the holding cages in the interior, and this same type of door also secures each end of the hallway connecting the exhibit to the holding area. These doors slide up and down and have a barrier of thick wire mesh. Using pneumatic pressure, an operator can control these doors from a central panel in a room on the level above the holding area. The room has a Plexiglas floor so that the keepers can see what is going on in the area below. The other doors leading to stairwells, etc., are also formidable. These "jailer" doors, made of reinforced steel, are the kind of barriers you might see in a *60 Minutes* documentary where the warden pulls out a chain of big brass keys to let Mike Wallace in to interview a particularly dangerous criminal.

Orangutans are extremely strong, and to prevent the apes from forcing open the pneumatic doors, the designers added extra latches on the top of the door that could be secured by pins with lugs on them. When the door is closed, the top of the door fits between two plates that can be accessed from the second level. A keeper then inserts the pin through keyhole-like apertures in the plates and in the top of the doors. The five-inch pin is then flopped over so that it cannot be withdrawn without flipping the pins into the proper position. Taken together, these various redundant security systems should have been able to contain most humans, much less an ape that rarely uses tools in the wild. But back in 1986 Jonathan had a motivation that went beyond his ordinary desire to escape just to prove that he could.

One of the less chivalrous aspects of orangutan society seems to be a tendency among males to force themselves sexually on unwilling females. Jonathan did this to show his dominance to human visitors, and according to one keeper, this caused the females to cringe whenever a new person showed up in the interior of the exhibit. In any event, Jonathan seemed to take this behavior as a prerogative, and he was not happy when he was denied access to female orangutan companionship.

Thus it was with some trepidation that the keepers isolated Jonathan from the main group after one of the females became pregnant. It was done for the safety of the mother, but Jonathan, confined to a holding cage in the interior, perceived this as an injustice and a challenge. The keepers tried to make his solitary confinement a little more tolerable, giving the orangutan a big truck tire and other toys to play with. And Jonathan was not alone in the holding area. In an incubator in a service area was Jonathan's infant son Joseph, then one week old, while his baby daughter, little Rudi, occupied a play area in another cage in the holding area. At various times, volunteers would come in to spend time with Rudi. Because Jonathan was secured behind pneumatic doors, the docents would leave open both the pneumatic door to the hallway and the jailer doors leading to the service area and to the lower level.

Jonathan at first tried a number of different ways to defeat the security systems. He would put hay at the bottom of his cage door to prevent it from closing all the way, then use his amazing strength to try to push it up. A keeper would then jam the pin through the mesh of the door to block it from opening farther. On more than one occasion, when the keeper came downstairs to remove the blockage, Jonathan had hidden himself and would try to grab the keeper's ankle.

Then Jonathan tried a new strategy. One of the docents who was a regular visitor began reporting to the keepers that sometimes when she came to look in on Rudi, Jonathan would be standing on the truck tire, fiddling with something at the top of his cage. Geoff Creswell, at that time a junior keeper, also regularly visited Jonathan, but whenever he came in, the orangutan would be sitting quietly by himself in a corner of the cage. At this point, Creswell and Jonathan had an uncomfortable relationship. Geoff says that Jonathan may have held a grudge because Creswell had begun working in the exhibit right around the time Jonathan was isolated from the females.

The purpose of Jonathan's mysterious fiddlings became clear one day as Creswell was returning from his lunch break. Walking through the hallway that linked the exhibit and the holding cages, he passed the service area. On his way by, he glanced through the glass on the jailer door and noticed that something was amiss. Someone or something had moved the incubator which held the newborn Joseph. Creswell decided to investigate and entered the service area. As he pondered the mystery of the moved incubator, his ruminations were interrupted by a noise. When he lifted his head, he saw Jonathan coming at him from the holding area.

An adult male orangutan looks gigantic, but an adult male with hair on end and huge arms outstretched looks like King Kong. It is easy to see why some people who have stumbled on orangutans in the wild have described them as ten feet tall.

Geoff did not wait around to take measurements, but hastily dove through the jailer door that separated the service area from the kitchen and slammed it behind him. Jonathan came up to the door, braced his feet against it, and started to pull the door handle through the steel door. While he did not open it, he did manage to bend the steel. Turning his attention to the service area, he spied the incubator. (Geoff is convinced that Jonathan did not know that a baby orangutan was in the incubator because he did nothing to harm Rudi, who sat vulnerable in her open cage throughout Jonathan's marauding.) Jonathan picked up the solid metal-framed incubator by two of its legs and launched it down the stairwell. (Secure within the incubator, Joseph survived this unscheduled flight with no ill effects and went on to lead his own life of crime.)

By this time, Creswell had found a radio and sent out an alert. "Usually when this happened," he recalls, "half the staff would come running and the other half would head for the hills." When Jonathan saw the arriving posse, he became even more agitated and started whooping and slamming one of the steel jailer doors back and forth. In the meantime, Geoff made it up to the control area for the pneumatic doors. Through cajolery and perhaps because Jonathan knew that the jig was up, the keepers eventually coaxed the orangutan back into a relatively confined area next to a steel holding cage. Then the zoo veterinarian was lowered by rope into the holding cage in the service area, where he shot the agitated ape with a tranquilizer dart.

Upstairs at the controls of the pneumatic doors, Creswell noticed that the pin that secured the pneumatic door from the top was lying on the Plexiglas floor. Later, reconstructing the escape, Geoff and the other keepers figured out that Jonathan had used a piece of cardboard to reach up to the pin and, by jiggling it, eventually get it in position so that he could back it out with the same piece of cardboard. Jonathan made it easy for the keepers to reconstruct how he pulled off his escape because as soon as he was back in the cage, he started trying to back the pin out again.

By watching his keepers, he seems to have figured out how the doors were controlled and how the latch was secured. Geoff also surmised that Jonathan made a distinction between the ranks of the various keepers. He made only partial efforts to conceal his fiddlings from the docent, but he would not work on the lock when Creswell or the other keepers were around.

To prevent new escape attempts, the keepers wedged a broomstick against the pin so that Jonathan could not flip it into the position necessary to back it out. Creswell reports that three years later the keepers decided to see what would happen if they took the broomstick away. Wasting no time, Jonathan immediately began working on the locking pin again.

Jonathan staged a number of other escapes whenever the opportunity presented itself. On one occasion, he got out of the holding cage and tried to use a ladder left lying around as a battering ram against a weak point in a jailer door. Another time, he took advantage of a door accidentally left open. When keepers found him, he was in the central area. He had ripped a hose out of the wall and was filling cups with water, which he was then pouring into a bucket. Jonathan next tried to put on boots and Playtex gloves, and, says Creswell, he picked up a squeegee and did his imitation of a workman cleaning the floor.

While a comparative psychologist might cite this last episode as an example of blind mimicry, the planning, tool design, and reverse engineering that went into Jonathan's escape called on most of the abilities associated with awareness, tool creation, and metacognition. Kanzi, the pygmy chimp, may have found a rock and cracked it to make a knife, as described in the previous chapter, but he was told to go outside and find a rock. In Jonathan's case, his toolmaking took place despite the efforts of the humans around him, who were doing everything possible to discourage him from applying his toolmaking gifts.

Although the evidence is inconclusive, it also appears that the ape hid his efforts from those keepers he deemed hostile to his efforts, but not from volunteers he deemed harmless, perhaps displaying the type of awareness of the mental states of others that gives humans such an advantage. His mistake was that he did not seem to realize that lower-ranking people could still communicate detailed information to higher-ranking people—an error typical of three-year-old human children, but which they cease to make as they get older.

Through observation Jonathan figured out the crucial elements of the door-locking system and how the various components worked. Then he was able to use a tool he had fashioned by rolling and twisting a piece of cardboard (at times he also twisted bunches of hay into a type of stick) to flip the pin and back it out. Given the opportunity three years later, he tried again to knock the pin out. He pulled off both these feats with absolutely no encouragement from his keepers. This seems as clear an example of program-level imitation as any situation that might be confected in a controlled study.

While Jonathan's escapes are well documented, there is no indication that he is an exceptional orangutan. Fu Manchu, the Sumatran orangutan who used a piece of wire to effect a series of nighttime escapes at the Omaha Zoo, also cloaked his actions, found and fashioned a tool, and used his powers of observation to program a series of movements that would lead to temporary freedom.

This escape has attained the level of a zoo legend in the three decades since it happened (Fu Manchu died two years ago at the Gladys Porter Zoo in Brownsville, Texas). According to Jerry Stones, the head curator at Brownsville,

Fu Manchu, like Jonathan, was single-minded in his pursuit of his vocation as an escape artist. Whenever he entered a new exhibit, he would walk over and inspect the padlocks to see whether they had been left unlocked.

Fu Manchu's first series of escapes took place in October of 1968, when the zoo was closed to the public for the season. When the weather was good, the orangs would be let in to the larger outdoor enclosure for some fresh air. One such day, while working in another part of the zoo, Stones was approached by a breathless keeper who told him that the orangutans were in the trees near the elephant barn. Stones climbed up into the trees and coaxed the group down and back into their cages. He then checked the enclosure and discovered an open door that allowed access from the moat that encircled the exhibit to the furnace room. The orangs had gone through the door and then up a set of stairs that led to the great outdoors, pausing along the way to fiddle with the wiring for the furnace.

Stones chewed out the keepers for carelessness, and the incident was forgotten. Confident that the escape was the result of human error, the keepers let them again into the outdoor enclosure on the next nice day. Once again they escaped. This time, says Stones, "I was getting ready to fire someone."

Then, the next time they were let out, a keeper came running up to Stones, saying, "You've got to see what Fu Manchu's doing." With the keeper, Stones went back and positioned himself on a hill in a spot beyond the apes' line of sight.

As Stones watched, the orangutans entered the outdoor exhibit. Fu Manchu quickly climbed into the moat. Taking hold of the bottom of the door with one hand, the orangutan produced a stiff piece of wire with his other hand. Sliding the wire between the door and the casing, he slipped the latch and the door popped open. Once again Fu Manchu was out, followed by his family.

In short order, the renegade orangutans were recaptured. Stones now knew how the orangutan was getting out, but he had no idea where Fu Manchu kept the wire between escapes. Only a day later, as Stones was leading the orangutan back in from the outdoor exhibit, did he notice something shiny sticking slightly out of the orangutan's mouth. He went up to Fu Manchu, pulled back his lip (their relationship was solid enough that Fu Manchu would tolerate such poking and prying), and discovered the piece of wire. Fu Manchu had bent the wire into a shape allowing him to fit it comfortably between his lip and his gum. Apparently, the orang had been carrying the wire around in his mouth for some time, waiting for opportunities to present themselves. The case of the disappearing orangutans got a good deal of publicity in the press. For his ingenuity, Fu Manchu was made an honorary member of the American Association of Locksmiths.

Stones and Fu Manchu later moved to the Brownsville zoo. Here, too, orangutans proved their mettle in their efforts to defeat the security arrangements. This exhibit consisted of an artificial island bounded by a sturdy electrical fence. Ichabod, one of the male orangutans, figured out that while he would get shocked if he touched the fence while standing on the ground, he would not get stung if he jumped onto the wire and rolled over the fence without being grounded. The island also had a palm tree protected by an electrical fence. Stones reports that the keepers had to continually experiment with configurations of the apparatus to stay ahead of orangutan attempts to short-circuit the hot wire. After one female broke a hoop of electrified wire by cracking its porcelain insulator with a rock and shorting the wire by pushing it to the ground, the zoo engineers put up wires connected by insulators to the tree itself. One female doubled a sturdy palm frond around the insulator and, pulling on both ends, ripped it from the tree—again providing a fine example of toolmaking.

There are many other celebrated orang Houdinis. At the San Diego Zoo, a serial orangutan escape artist named Ken Allen achieved such celebrity that an enterprising observer emblazoned a sweatshirt with headlines detailing his various breakouts. During one of his escapes, he propped bamboo against the wall and used it as a ladder to get out of his exhibit, a simple strategy that has been used by all manner of animals.

Marvin Jones, now retired, was the registrar at the San Diego Zoo, and for fifteen years he kept the stud book which tracked the genealogies of all the captive orangutans held by accredited zoos in the world. He was at the San Diego Zoo during Ken Allen's series of escapes, and remembers well the brouhaha they caused. The escapes began right after the completion of a new facility for the orangutans that put the animals in side-by-side enclosures with jungle gyms and trees in the center of the separate exhibits.

When Ken Allen was put in the exhibit, he began inspecting it for possibilities. "Invariably an orangutan will find the weakest part of an exhibit and begin working on it," says Jones. Although most of his attempts went unobserved, it appears that Ken Allen went down into the moat and discovered a way to inch up the wall. "Each of an orangutan's fingers is like a human hand in terms of strength," says Jones, "and so it's as though they have five hands on each arm, and with that kind of gripping power, they can climb a lot of things you would not think can be climbed."

After escaping, Ken Allen tended to walk around to the front of the exhibit and simply sit down. He had been hand-reared and so was used to people, and more than one visitor came up to him to pet him during these escapes. Once he was caught, the keepers simply walked him back to the enclosure. "I don't think he was trying to get away," says Jones. "I think he was just finding a problem to occupy his engineer's brain." The zoo adjusted the design of the moat, and Ken Allen's escape attempts stopped. Now in his twenties, he is still at the zoo, but has settled into a more sedentary lifestyle.

Another orangutan escape that involved deception and long-term planning also served to underscore the point that breakout attempts do not necessarily mean that an animal is dissatisfied with its living conditions. When he was a teenager in the late 1980s, Chantek, the orangutan who was taught sign language by Lyn Miles, staged his own break-out from his room in his trailer in Chattanooga, Tennessee. Diligently working over a span of days (Miles is not sure how many since she only discovered the deception later), Chantek gradually unraveled the chain-link wire that separated his room from the rest of the house trailer, but in just the moment before Lyn or her assistants arrived again, he would rearrange the wire to conceal his tampering. Once he had created a big enough hole, he broke out and trashed the trailer.

On the other hand, there is also strong evidence that Chantek saw the trailer as a refuge. When he was a juvenile, Ringling Bros. and Barnum & Bailey Circus came through town, and its managers offered Lyn a chance to bring Chantek to meet the animals. All went well until the orangutan was offered a chance to pet a tiger. Freaking out, Chantek jumped out of Lyn's arms and ran through the University of Tennessee campus, across the football field, and back into his trailer.

Once inside, he flipped over a sign that Lyn had created for the trailer window which said that she and Chantek were out but would return soon, in effect telling his imaginary pursuing tigers that Chantek was not at home. He then locked two sets of doors, forcing Lyn, who was breathlessly trying to catch up with her panicked charge, to come in through the window. As a final precaution, Chantek got in his sleeping hammock and pulled the covers over his head.

Lyn says that it took a day of coaxing to get him to come out of hiding. From that point on, as part of Chantek's bed-time ritual, Lyn would carefully look out of the trailer win-dows and say that there were no "cats" outside. The trailer incident provided Lyn with a weapon on those occasions when Chantek refused to come inside. All it took was a men-tion that "bad cats" were in the area, and Chantek would come flying to her side. (I should mention that Chantek liked house cats, once adopting a stray whom he named Kitty by making the sign-language symbol for "cat" with each hand.)

Unlike other orangutans, Chantek could draw on language skills in his efforts to get out of an exhibit. After he had spent fourteen years with Lyn, the University of Tennessee returned Chantek to the Yerkes Regional Primate Center, and for several years she was not allowed to see the orangutan. Responding to her pleas, Terry Maple, the director of ZooAtlanta, arranged for Chantek to be brought to the zoo, and Lyn had a reunion with the red ape that was filmed by ABC television. At first Chantek seemed shaken to have Lyn reappear in his life, she says, but after a while he settled down and signed "ice cream," one of his favorite treats. (Chantek was the ape who slimmed down from over 400 pounds to 245.) Then Chantek's thoughts turned to getting out of his new cage.

First he went to a door and made the sign for "open." Lyn signed "No." Then he signed, "Get key." Lyn told him she couldn't. Then he pointed to one lock and signed, "Get car." This time Lyn signed, "Go where?" Chantek replied, "Go home." Lyn signed, "This is home." Chantek not only remembered his sign-language vocabulary but also methodically exploited the language to try to get out of the zoo.

While orangutans are the superstars of zoo escapes, other apes have used props and tools to get out of exhibits. At the Woodland Park Zoo, Kiki the gorilla propped a log against the moat and climbed out. Once out, she assumed the role of a zoo visitor. When the keepers found her, she was sitting in the bushes above the exhibit watching the polar bears in a neighboring enclosure. (It is a safe assumption that gorillas never get to see polar bears in the wild.)

There are also several stories of gorillas, chimps, and orangutans cooperating in escape attempts. At the Arnhem zoo in the Netherlands, a group of chimps formed a pongid pyramid so that one could get to the top of a fence. The first escapee then reached down and helped the others out.

Apes dominate the lore of escape attempts, but numerous

other species test the security of cages and exhibit areas. There is at least one story of a parrot showing some ingenuity in escaping from its cage and then facilitating a group jail break. According to Sally Blanchard, this episode took place in Wichita, Kansas, roughly twenty years ago in a breeding flock of about thirty double yellow-headed parrots. The escape took place while Bill and Wilma Fisher, the couple that raised the birds, were away at a parrot show. Chango, one of the parrots, used his beak to unscrew the bolts on his cage to the point where it collapsed. Once out, he somehow unlatched the other cages one by one. Sally Blanchard was looking out for the birds while the Fishers were away, and when she stopped by in the evening, almost all the birds were out and having a parrot party.

Nor are apes the only animals that have used props to effect escapes. Bonnie Beaver, an animal behaviorist at Texas A&M University, recalls that several years ago a family at the university discovered that their dog kept escaping from their backyard, which was enclosed by a high fence. Eventually the owners figured out that the dog was climbing up the woodpile and over the fence. The owners moved the woodpile to the center of the yard. The dog decided that this would not stand, and began dragging wood back to the fence to create his own pile.

This dog may have wanted to escape simply because dogs like to roam around. The different escape attempts I investigated involving apes were motivated by a variety of factors. Sometimes, as in the case of Jumoke, the Woodland Park gorilla, the animal wants to rejoin family and friends elsewhere in the zoo. But with many of the orangutan attempts, something even more subtle may be at work. An animal smart enough to figure out how to defeat a sophisticated lock, like Fu Manchu or Jonathan, is also smart enough to figure out that Brownsville, Texas, and Topeka, Kansas, are not orangutan-friendly places. They know that the zoo enclosure offers food and security, and it is safe to assume that they know that their human jailers are well-meaning. After the first escape, they also know that the worst that can happen (so long as they don't venture too far) is the indignity of getting shot with a tranquilizer dart. Against these costs, there are the immeasurable benefits of testing their wits against their human captors, and, for a few glorious moments, of watching the humans around them scramble in a panic.

Zoo designers have acknowledged the special gifts that orangutans bring to escape attempts by occasionally employing the apes as consultants. When the Los Angeles Zoo inaugurated a new chimp enclosure, one of the ways they tested its security was to bring in an orangutan and see whether it could escape. The zoo officials reasoned that chimps would not succeed in escaping if an orangutan failed.

As in the case of games, tool use, and deception, the orangutan's fateful attraction to escape may reveal something about the evolution of intelligence. Asked what it is that enables orangutans to escape where other animals cannot, different keepers and specialists give different answers. Some talk about their combination of insight and strength—an orang can unfasten a large screw with its fingernail—others, their fascination with machines, while Marvin Jones says it is their patience. "An orangutan will work and work at something until they get it. Chimps don't have that attention span."

He may be on to something. Having spent a good deal of time observing chimps when I wrote *Apes, Men, and Language*, I began thinking that access to one's higher mental abilities is almost as important as intelligence itself. It is possible to be conscious but, as Sally Boysen's and Marc Hauser's work implies, too caught up in events to make complex assessments of the world around you. For very good reasons, nature greased the lines between perception and action in both predator and prey species of all types. The ability to stand back and assess before acting comes with the price of increased risk. The more highly social an animal, the better it is able to rely on the eyes and ears of others in its group, and the lower the risks of temporarily ignoring the continuous stream of information being delivered by its senses.

At least this is one cost-benefit analysis implicit in the evolutionary history of humans and some other highly so-cial animals that nature has equipped with the capacity to test hypotheses in the mind before implementing them in the real world. As noted in the discussion of Sally Boysen's experiment in an earlier chapter, only when centers in the neocortex develop sufficiently can a human or animal over-rule the simplistic urge to point at what they want, and in-stead figure out that the path to a larger portion of treats leads through pointing at the smaller.

Orangutans seem relatively well equipped to overrule these urges, although whether that means they have cortical-to-cortical connections in their brain that other apes have not yet developed is an open question. Saying that an animal has patience is saying that an animal has the chemical signals and neurological wiring to support patience. Attention span is not simply a function of temperament—otherwise cows would be way ahead of us.

Still, time, in the sense of access to some mental workspace, is one of the dimensions necessary to manipulate the world in the mind. This mental working space can only be as elaborate as the time an animal has to divert its attention from the world and enter this reified world of images, symbols, and the rules and relationships that govern them. This is what is im-pressive about the more elaborate escape attempts executed by Jonathan and Fu Manchu. Each involved a lot of moving parts and opportunities for failure. From finding materials that might be made into tools to concealing their intentions and activities to choosing their opportunities for escape, the orangutans had to keep a lot in their heads. More than almost any other activity of captive animals, these escape attempts make me want to know what else is going on in their heads.

LOVE! VALOR! COMPASSION!

Empathy and Heroism

N OT every animal can match the orangutan for its planning skills and dexterity or the gorilla for its humor, but all manner of species great and small have demonstrated the traits which we cherish most in humans: loyalty, love, trust, and heroism. It is, of course, far easier to adduce awareness to some Machiavellian bit of deception than to an act of love or trust. After all, the more intelligent lemmings would be the ones who did not follow the leader off the proverbial cliff, not their dim-bulb brethren who plunged ahead.

Still, trust and awareness do converge at times. I've already cited a few examples. When Bonnie, the orangutan at the National Zoo, brought her infant Kiko up to the bars so that Rob Shumaker could give the sick little ape an injection, it revealed not only extraordinary trust, since orangutans hate injections, but a judgment that the baby needed care beyond what she could provide. Bonnie also trusted that Rob might be able to provide that care. To bring Kiko up to the bars, Bonnie had to overrule one of the most powerful drives in nature, the urge of a mother to protect her ailing infant.

A more humorous version of this combination of trust and awareness took place at the Woodland Park Zoo in Seattle.

Helen Shewman says that on one occasion the orangutan Towan escaped and keepers were forced to shoot him with a tranquilizer dart. Towan was one smart cookie, however, and having had some experience with being darted, he figured out how to pull the dose of tranquilizer out of the dart/syringe that delivers the drug. The keepers managed to lure Towan back into his cage, but because he had never been knocked out, they weren't able to extract the syringe from his shoulder.

For some time he wandered around with the barbed needle dangling and he did not like it. He tried to pull it out himself, but it clearly hurt, and so he eventually came up to the bars and put his shoulder against them so that Helen could try to pull it out. It was stuck too firmly for Helen to get a good grip, and so she went away to get a tool. When she returned, she showed Towan a pair of pliers and told him that she was going to have to use them to pull out the dart, and as she talked she used the pliers to grip skin on her own arm to show Towan how she was going to use the tool.

At first Towan wanted nothing to do with what looked like an instrument of torture, but after pacing around a bit, he seemed to arrive at a decision. He came up to the bars and pressed his shoulder against them so that Helen could reach the dart. Then, in the manner of any child who has ever been scared of an injection, he turned his head away and covered his eyes with his other arm while Helen pulled out the dart. I'm sure that Towan appreciated Helen's ministrations, but I'm equally sure that he will try to escape again.

Harriet, the leopard cited in the introduction, also seemed to demonstrate trust and awareness. She made the judgment that Billy Singh, the Indian conservationist who raised her, could provide a more secure ride across a raging river than she could herself. Like everyone else who has seen that wildlife film, I was impressed by the practicality of her maternal instincts—moving her cubs onto the very high ground of Singh's kitchen, and then calling on him to get her back across the river. But there are other aspects of that story that are also striking. Harriet recognized that the water was receding and that it was safe for her to take her cubs back across the river. This in turn suggests that she had some sense that if the water was receding on Singh's side of the river, it was also receding at the den. Perhaps she had some mental map of the region. She certainly seems to have had a sense of the relative elevation of different places in her range.

Captivity (and domestication) can change the rules of who is friend and who is foe. The wolf and goat that became friends at the San Diego Zoo are a good example of the strange friendships that can arise. Maureen Fredrickson of the Delta Society has an Amazon parrot that became friends with a cat. Both shared a fondness for cheese, and the parrot would hold the cheese in its foot for the cat to eat.

On her own ranch, Maureen witnessed one of the strangest alliances imaginable. One day she went outside and saw that a wild turkey was standing on top of her horse Murphy. Thinking it was an accident, she put the incident out of her head. But the visits continued, and one day she saw the turkey arrive and the horse lower its head so that the turkey could climb aboard. Maureen reports that this went on for about a year. Maybe the horse, a less than successful racehorse, was thinking, "If I'm stuck being ridden, a turkey is sure a lot lighter than a person."

These strange friendships do not require intelligence, but they bespeak a certain flexibility. If we readily accept that animals can form bonds of friendship with humans, who in the course of things have eaten or continue to eat most of the creatures we cultivate as companions, then it should not be so strange that animals might bond with other odd bedfellows as well. While scientists might debate the degree to which cross-species bonds are accompanied by awareness—does the Cretan goat at the San Diego Zoo marvel at the fact that it has made friends with an animal that would ordinarily eat him?—there is no ambiguity about the risks animals have taken to protect those they see as family.

Animals are even willing to take big risks to protect creatures they don't particularly like if they are members of their extended family. Back in the 1970s, my then mother-in-law had two miniature poodles, which she doted on. Their frenetic natures and high-pitched barking drove animal-loving humans and, all the more, my cat, Zephyr, to distraction. Since all the animals in this story are long since deceased, it is possible to write about the particulars without offending human or four-legged sensibilities, and I can say that Zephyr really could not stand the generations of small poodles he encountered during visits. The feelings were reciprocated, and once, after Putzel, the matriarch of the poodle clan, had a litter, Anne Gill placed a sign at cat-eye level on the door of the room that served as a nursery that read PUPPIES INSIDE, DO NOT DISTURB! ZEPHYR, THIS MEANS YOU!

Earlier that year, when Putzel was in heat, all manner of neighborhood dogs began sniffing around the yard of the house in Bronxville hoping to get lucky. On one memorable day, a big black Labrador had tiny, terrified Putzel cornered and whimpering. As I came out to see what the commotion was, who should come galloping up to launch a ferocious attack and drive off the Lab but Zephyr. It's possible that Zephyr, a tom, was defending his territory, and I have to say that he didn't hang around to console the beleaguered poodle. Still, I'd like to think the incident revealed a little feline gallantry.

Heroism is a concept that has become much degraded in its application to humans. Politicians and the media use it almost interchangeably as a label when they really mean "victim." Putting aside its overuse in this hyperbolic age, heroic acts of physical bravery mean an individual overcomes any fear and risks his or her own life for the sake of principle or others. Apart from the issue of standing up for a principle, I would argue that selfless acts of physical courage come from an ancient response that predates the species, one shared with many other animals. Therefore it is only natural to assume that other creatures are also capable of heroic acts, which is to say that other creatures that put themselves in harm's way for the sake of others have to overrule the same instincts for flight and panic that we do.

Most people recognize this, and virtually every state in the union gives awards for pet or animal heroism. Ken-L Ration has its Dog Hero awards, and the acts speak for themselves. Some years back the winner was a dog named Bo. Along with another dog, he accompanied his owners on a white-water trip down the Colorado River. In a particularly turbulent stretch of water, the raft flipped over, tossing the husband and one dog free, but trapping Bo and the wife under the overturned raft. Bo quickly popped up, but then dove back under the raft and, grabbing the woman's vest in his jaws, dragged her out.

There are countless similar stories—a Rottweiler dragging an injured woman from a burning van; Elizabeth, the cat who would come running in to save her veterinarian owner from fractious pets she was treating; Shade, the cat who stood between Nell Mitchell, a bedridden woman, and her attackers. Between books like *Real Animal Heroes*, pet hero awards, and periodic heartwarming accounts in *People* magazine, these acts of courage are sufficiently well covered that I will not dwell on them here beyond retelling the tale of a courageous pig.

In 1984, Priscilla won the American Humane Society's Stillman Award and was made the first member of the Texas Animal Hall of Fame for her efforts to save a small child from drowning. Priscilla was with her owner, Victoria Herberta, one summer day at a park on Lake Summerville, which lies north of Houston. Fond of swimming, the pig was paddling happily around the lake with Victoria's friend Carol Burk and her mentally handicapped son. The three-month-old pig had started in toward shore when the eleven-year-old boy got in trouble farther out. Priscilla was the closest mammal to the child, and the mother screamed for Anthony to grab the harness the pig always wore. The boy grabbed the pig's harness, and the forty-five-pound pig towed the eighty-five-pound boy 150 feet, until the boy's feet could touch bottom.

One of the most moving and celebrated stories of animal heroism involved a stray cat in New York who rescued all five of her kittens from a burning building, suffering disfiguring burns in the process. The incident was reported by Dave Gianelli, an animal-loving fireman who heard kittens meowing during the height of a blaze at an abandoned auto shop. He knew someone was rescuing the kittens because the number of kittens on the sidewalk kept growing until it reached five. Then he discovered the mother, shaken and badly burned not far away.

Gianelli took the cat family to Long Island's North Shore Animal League. There the vets were astonished that the stray female cat tolerated their ministrations. She seemed to know that people wanted to help her. The story touched a tender spot in both the media and the public. The league was getting daily calls from journalists as far away as Paris and London, and thousands of offers poured in from people who wanted to adopt the cat and kittens. Scarlett, as the stray was dubbed, was reunited with her kittens long enough for her to know that her heroism had saved their lives, and then the cats went off to their adoptive homes. Scarlett now lives in Brooklyn and still gets regular visits from the fireman who saved her life.

If an animal bonds with a human, it is not unreasonable to assume that it shares at least some of the emotions that, in humans, accompany such bonds. A loyal dog in the eyes of a behavioral scientist may simply be transferring bonds associated with a pack to humans, but the dog is not a windup toy, and it is likely that those bonds are supported by emotions.

Perhaps because death is so fundamental to life and because the death of a member of one's group is felt so acutely among social animals, grief has a language that is easily understood across the species barrier. Maureen Fredrickson spoke of a dog that, as part of a program to enrich the lives of seniors confined to nursing homes, visited an aged woman until the day she died. Because the woman had become so devoted to the dog, the family invited the dog and its human family to the funeral and wake. After being led to the coffin, the dog was taken outside. According to Maureen, it stood back and howled four times. Then it walked away. Who could want a more eloquent requiem?

There is no question that any number of other animals can, through smell and sight, tell when a quickening heartbeat has stopped. It is also possible that they can sense when someone is about to die. At a home for the aged in Oswego, New York, a cat named Smoky is one of the pets that roam the facility, giving comfort. As reported in the documentary *Extraordinary Cats* produced by the PBS series *Nature*, Smoky seems to sense when the patients are in the final hours, and he will seek them out and lie with them.

There is no doubt that any number of animals are affected by the death of an offspring or member of their group. In elephant groups, mourning takes on almost ritualistic overtones as different members of a matriarchal clan console a mother who has lost an infant, or those elephants closest to an adult that dies. Elephants will inspect the bones of long-dead members of a group years after the animal has died. Perhaps some minute chemical signal lingers in the bones of the dead that summons elephant memories in those that visit the bleached remains.

The question is what form grief takes, not whether it exists. I would not at all be surprised if elephants, who live in highly complex and large social groups, had a degree of awareness that allowed them to imagine the finality of death and ponder the meaning of mortality. Perhaps orcas can do this as well.

The story cited in the introduction in which the killer whale Orky slammed his head against the wall just before his mate, Corky, miscarried suggests not only the degree of grief an orca might feel, but also that the marine mammal was aware that Corky was carrying his offspring. What could have been going on in Orky's mind at that point? Gail believes that Orky was monitoring the pregnancy, using his ability to image with sound. Before this pregnancy, Corky had already lost several previous infants. Here it is perhaps reasonable to anthropomorphize, and assume that Orky's grief was also exacerbated by extreme frustration.

The fact that Orky and Corky mated at all suggests that captivity was tolerable enough for them to follow life's imperative. According to Gail, everything about Orky's behavior suggested that he desperately wanted a baby, right down to the fact that he let Corky boss him around when she was pregnant. As described earlier, he even cooperated with humans when another of their offspring was sick, offering his body as a platform from which a worker could adjust a sling that was being used to carry his baby out of the tank for treatment. It is easy to understand that Corky would grieve the miscarriage as any mother would, but Orky's pain suggests a combination of anguish, awareness, and happiness denied with which any human can sympathize.

Grief is probably universal among all sentient creatures. Empathy is another matter. Unlike grief, it requires that one creature understand what another creature is feeling—and that it care. On the first score, there is plenty of evidence that many animals have an ability to pick up on moods and feelings that approaches the paranormal.

Indeed, much of the work of the Delta Society, which fosters the use of animals "as healers, allies, and friends," is predicated on the extraordinary sensitivity of animal senses shaped by the needs of survival. Dogs, whose wild canid ancestors had to be acutely aware of the mood of other members of their group, have been used in the therapy of battered children. They will often seek out and approach the child who is suffering the most severely at any given time.

Maureen Frederickson also tells of dogs who can sense when diabetics are about to have a seizure. On one occasion, a dog kept knocking a woman down (as she was trained to do) to prevent her from leaving her house. Not long thereafter, the woman went into a seizure. Somehow, dogs seem to be able to smell low blood sugar.

Dogs also seem to be able to sense physical weaknesses not obvious to the human eye. There are reports that they can sense the location of tumors. One dog, named Joey, was assigned to a woman who had multiple sclerosis, a degenerative disease of the nervous system. On certain days the woman was less stable than others. The dog not only sensed when the woman needed support but would also walk on the woman's left side, her weaker side, and lean supportively against her leg while they walked so the woman's knee would not buckle. This had not been part of the dog's original training.

Parrots also have an ability to sense physical change in their environment that borders on the eerie. A client of Sally Blanchard's in San Francisco had a parrot who would say "Dad's home" a few minutes before the father would enter the house. The parrot's owners were mystified as to how the bird knew this and launched their own investigation. When the parrot said "Dad's home," the wife called her husband on the cell phone to see where he was. It turned out that the husband was a few blocks away, and the couple surmised that somehow the parrot recognized the particular sound of the owner's car as he downshifted at that point to turn up a big hill.

Blanchard says that parrots and a number of other birds

have cells on their ankles, knees, and leg joints called Herbst's corpuscles that seem to serve as very sensitive vibration detectors—a very useful piece of survival gear if you are a one-pound bird sitting in a tree surrounded by a host of raptors, mammals, and reptiles that would love to make a meal of you. Blanchard and others speculate that these same cells provide the early-warning system that enables them to predict earthquakes. She reports that her birds started screaming about fifteen minutes before the big San Francisco quake of 1989, and that after the quake she had a flood of calls from parrot owners saying that their birds had begun freaking before the trembler hit. (Other animals also seem to sense earthquakes. The number of animals reported missing doubled in the days before the October 13 quake, suggesting that pets were heading for the hills for safety.)

Layne Dicker says that when he is eating pizza, Chicken, his thirty-year-old Amazon parrot, watches him closely. He says that at the exact point he begins thinking about giving Chicken a piece, but before he has begun to break off a piece, the parrot will say, "Oh boy!" Dicker attributes this not to mind reading, but to the extreme powers of observation of a prey animal.

Reading moods and feelings is one thing; caring about how another feels is another. Whether they empathize or not, parrots, at least, can use the human language of caring appropriately. Sally Blanchard says that after one of her clients learned that she had a bad disease, she walked into the living room, whereupon her parrot asked, "Are you OK?" It was something that the woman said to the parrot all the time, but this was the first time the parrot had used the words with her. Sally herself once had a bad headache, and said to her parrot, Paco, "I have a headache, so you have to be quiet." Paco simply said, "OK."

It is somewhat unnerving to hear parrots using the language of caring with each other. When I first entered Sally's house, a chorus of little voices erupted. From the other room

I heard a bird say, "Hi, Sawie" (parrots have a hard time with *l*'s and *r*'s). In the room were Paco and Rascal, who chattered away in English to each other. Among the phrases they use are: "Oh, you're so pretty," and "I love you." Since parrots are so highly attuned to emotions, it is entirely possible that the birds were intentionally conveying the sentiments of the phrases even though they might not know the exact meaning of the words.

Maureen Fredrickson has no doubt that animals are capable of caring about us as well as sensing pain. She says that every day she encounters evidence of animals demonstrating affection and concern for humans they are asked to help. She cites numerous examples of dogs, cats, pigs, and horses going well beyond their original training to show sensitivity to injured psyches and injured bodies. In many cases, the animals' efforts are directed on behalf of people they scarcely know.

The skeptic would argue that the animals used in therapy have been trained through reinforcement and rewards for their roles. But as anyone who has dealt with the motor vehicles department or encountered Soviet-style bureaucracy knows, there is a world of difference between the rote performance of a task and cooperation energized by an emotional involvement on the part of the helper. The stories of dogs and cats and pigs and horses that make life easier for the wounded and afflicted are interesting and moving because they suggest a capacity for generosity and empathy in animals. This may not be as dazzling as evidence of higher mental abilities, but it is certainly as poignant. Moreover, these stories suggest (without proving) some degree of awareness of the mental states of those they comfort.

WHAT DO THEY MAKE OF US?

A Place Where Humans Are the Novelty

THIS book in large part has been about the ways in which animals bring their mental abilities to bear in their dealings with humans. Because we dominate the planet, and because this supremacy is a relatively recent event, most animals have not been prepared by evolution to cope with the ways in which we intrude into their lives. In their dealings with us, animals, particulary those in captivity, fall back on instinct and apply the lessons of trial and error. But they also draw on whatever higher mental abilities they possess.

I have focused on things animals do that reveal these abilities because it is far easier to reason about the meanings of actions than it is to ponder the contents of animal daydreams, solipsisms, or other private mental events. Much of our own mental life never translates into actions and thus remains inaccessible to other humans, and so, barring some breakthrough in the mapping of how thoughts stimulate precise electrical events in the brain, the interior lives of animals will remain in the province of speculation.

Still, there are things we would like to know about the mental life of animals. One of them is simply: What do animals make of us? How do *they* see humans? This question ventures into what is perhaps the most private and subjective realm of experience of all. Most of us spend a good deal of time successfully concealing our true feelings towards other people. Attempting to probe such feelings across the species barrier poses what most scientists would regard as an insurmountable array of pitfalls. Let's try anyway.

One way to dig into this question is to draw upon what we know about how animals see other animals—e.g., whether animals have a sense of where they fit in a hierarchy. Seyfarth and Cheney's work with baboons, mentioned earlier, suggests that the answer is yes, that at least some animals can look at a hierarchy and see how they fit into that hierarchy. If it is true for baboons, it is also likely to be true for a number of smarter social animals, including chimps, bonobos, and orangutans, as well as dolphins, orcas, perhaps parrots, and other highly social, highly intelligent animals. Some field researchers argue that the ways in which an animal sees itself may differ according to gender. In *Silent Thunder*, Katy Payne's account of her studies of elephant communication, she writes that if she were advancing arguments that elephants are self-conscious, "I'd suggest that male elephants see themselves as individuals, and females have a sense of themselves as members of a community."

If to some degree an animal understands a hierarchy and can see its position in that hierarchy, then, chances are, that same animal in captivity will become a student of the human hierarchy around it, and perhaps see itself in that hierarchy as well. Recall that Jonathan hid his preparations for escape from the keepers, but not from a volunteer at the Topeka Zoo, perhaps making distinctions in rank among the people he encountered regularly. (Perhaps he also categorized them in terms of their potential to thwart his plans.)

Intelligent animals recognize that their human keepers provide food and control the technologies that control their lives. So do cats and dogs to some degree. Every pet owner has been led to the food bowl when he or she is late delivering breakfast, and when my cats want to be let outside, they will run to the door and look back up at me, posing a question with their eyes. By Harriet the leopard's actions it is clear that when she decided to return her cubs to their den, she also recognized that Billy Singh was the key to a safe ferry ride across the swollen river.

While animals recognize that humans have special powers, they do not automatically cede all humans respect and authority. Chimps will challenge a visiting human every chance they get, as will big cats and innumerable other animals. As noted earlier, Orky demanded that humans treat him with respect, but on occasion he and Corky turned to humans for help in dealing with their babies. Tim Desmond recalls that one time, the two orcas were looking down from the surface at their baby, who was lying on the drain on the bottom of the tank. Growing alarmed at their listless offspring, they then picked their heads up and looked at Tim and the other trainers gathered at the top of the tank. "They were just kids themselves," recalls Tim, "who knew nothing about raising a baby, and their look was a clear question: 'What do we do now?' " Before the keepers dove in, the baby roused itself and came to the surface.

An animal may develop a close friendship with a human, one which reciprocities like any friendship. Wayne Williams, a keeper at the Woodland Park Zoo, shared his lunch with a hippo every day he was at the zoo. Obviously, any morsels the giant herbivore received had more symbolic than nutritional value. In return, the hippo gave Williams affection.

Sometimes animals decide that a particular human might make a good mate. A number of female dolphin trainers have had to deal with amorous male dolphins (one of the more oversexed animals in the ocean). It seems as though nature contrived to make humans as appealing to dolphins as they are to us. It also seems that orangutans, gorillas, and chimps at least recognize us as another primate since a number of male and female researchers have had to let various species of great ape down easy. It took a while for Penny Patterson to convince Koko that male gorillas were sexier than male humans. While Koko was never more than mildly flirtatious with me, from time to time I have had to deal with sexually emancipated female apes.

In the wild, all manner of animals learn to distingush between humans who pose a threat and those who do not. Unfortunately, throughout the world, vastly more humans pose a threat to animals than do not, and virtually every wild species with legs or wings will head for the hills at the sight, sound, or scent of people. Even animals that have learned to be wary of humans can become habituated to certain humans like gamekeepers and researchers and special places like parks. In some parts of Africa, vervet monkeys supposedly have a different alarm call for people with guns than they do for the unarmed.

Fear of humans, however, is a learned behavior. It is interesting that if a population of animals have no prior experience with people, these creatures' first instinct upon encountering a person is not to flee but to investigate. Moreover, in those rare places where humans have made a point of not hunting animals, that curiosity continues. In Antarctica, for instance, humans have left penguins sufficiently alone that they still regard humans as a form of entertainment rather than as a new predator. To wander among penguins at Cape Royde near McMurdo Sound is constantly to be approached by inquisitive penguins.

In those rare places on the planet where humans are still a novelty, visitors have the opportunity to experience what it is like to be the object of study. I had this experience in a remote and inaccessible part of the central African rain forest. The story is worth retelling here because it offered an exceedingly rare glimpse of nature as it is without humans in the equation, and, more pertinent to this book, it suggested some of the ways in which intelligence in many forms infuses the natural order.

The place is called the Ndoki, a 7-million-acre expanse of rain forest (about the size of Belgium) that lies east of the Sangha River, just below the border that separates the Central African Republic from the Congo. For more than a thousand years, this stretch of forest has been protected on three sides by enormous swamps, unnavigable rivers, and arid hills to its north. The name Ndoki means "sorcerer" in Lingala, the common language of French Africa, and the region has also been protected by myths and legends—at one point during my fifteen-day expedition through these forests, our Pygmy trackers and bearers refused to go on. Pygmies hunt in virtually every forest in Africa, but not here, and they told us they were afraid of encountering Mokele Mbembe, a dinosaurlike creature that, legend has it, inhabits these forests.

There is a good deal of other evidence that the Ndoki has remained inviolate for a long, long time. Throughout Africa, *Funtumia elastica*, a type of rubber tree, is marked with slashes, the result of Pygmies harvesting latex, but not here. Throughout equatorial Africa, the riverbeds contain palm nuts, evidence of prior human habitation, but not here (at least not in 1992, but more about this in a moment). And throughout Africa, the various Pygmy tribes who are the true forest-dwelling peoples have legends and songs about the terrain, but not here. Perhaps the most telling evidence of all is the behavior of the animals themselves. Throughout Africa, forest pigs, monkeys, and apes disappear at the first sign of human intrusion—but not here.

Only several years after I first heard of the Ndoki did I decide to visit the forest. Michael Fay, a botanist with New York's Wildlife Conservation Society, told me about this wondrous place when I met up with him in Brazzaville in the Congo in the late 1980s. Then, in July 1990 at the Thirteenth International Primate Conference in Nagoya, Japan, Masazumi Mitani again brought up this remarkable place where the animals do not run away from humans. Along with Suehisa Kuroda, Mitani had established a research camp on the edge of the vast area in 1987. While the place sounded like Eden, Mitani also said that it was "very, very difficult." Having witnessed the hardships that Japanese researchers endured without complaint at other stations in central Africa, I had to wonder what the area might hold that would impress a scientist who was regarded as a hard case even by other Japanese field biologists. I was to find out.

I'm not one of those writers who seek out discomfort for the sake of having a good story to write about, but it was not the aura of menace surrounding the Ndoki that at first dissuaded me from visiting—rather it was the fear that publicity might bring to the region the human contamination it had successfully avoided for at least a millennium. But then, in 1991, Fay told me that the Congo government had granted logging concessions to virtually every tract of forest surrounding the Ndoki, and that timber ventures had their eye on its trees as well. Now it became important to let the outside world know that something precious was in jeopardy, and so in June 1992 I set off with Fay; Karen Lotz, a photographer (and now the editor of this book); and seven porters and trackers, and headed east from Bomassa, a Pygmy village on the banks of the Sangha River, towards the vast unknown of the Ndoki.

I found out very quickly that the trip was not going to be easy. The fifteen-mile hike from Bomassa to the crossing point on the Ndoki River took one or two days in 1992, depending on how much the bearers had had to drink. (Throughout central Africa, Pygmy men tend to drink away their earnings, and though it may sound paternalistic, Mike Fay says that this helps preserve Pygmy culture because after a binge they are forced to return to the forest to earn more money.) We made the mistake of traveling ahead of our bearers, and our hung-over crew dragged its feet, forcing us to stop at the abandoned camp of Kuroda, the Japanese scien-

tist, near the Djeke River, a little more than ten miles outside
Bomassa. From bitter experience, Fay said that we could not
push the porters too hard or they would simply abandon us
in the middle of the forest.

The first day's hike was not too onerous, and so I settled
in to sleep wondering whether the dire reports I had heard
from the Japanese researchers had overstated the hardships.
A few minutes later, I awoke feeling an insect on my finger. Flicking it off, I felt another take its place, and then suddenly thousands of bugs seemed to bite me at once. Only
seconds later, I heard others cry out as they were attacked as
well. Stumbling blindly over roots and a massive column
of ants, we tore down a path and dove into the river. Crushing the ants seemed to release some chemical distress signal: as we emerged from the river, ants dropped on us from
everywhere.

Loath to bother Fay so early in the trip, I stamped and
slapped until exasperation overcame pride and I roused the
scientist, who was peacefully sleeping about fifteen yards
out of the path of the ants. Fay, however, seemed to have a
limited supply of empathy and advice. Surveying the insects
that still covered my legs, he said drowsily, "Driver ants can
really be a problem; they can kill a tethered goat." Then he
went back to sleep.

Left to my own devices, I set up a hammock away from
the column of ants. Distracted, I drove a spiky vine clear
through my thumb, causing a pulse of blood to spurt from the
wound. Then it began to rain, and at about 2:30 A.M. I heard
a leopard cough. Enough was enough; I decided to check out
my tent. The ants had gone, and I was able to grab a few
hours' sleep before dawn. It was a tough start, but it was
only much deeper into the trip that I discovered what Mitani
might have meant by "very, very difficult."

The next day we hit the swamps that have long deterred
those curious about the Ndoki. We picked our way through

the quicksandlike muck by feeling with our toes and walking sticks for a series of thin logs that Japanese researchers had previously laid down as a path to the river. I slipped once and sank chest-deep in mud before I arrested my plunge by grabbing a root.

The swamps gave way to the Ndoki River, which is the real barrier to the region. Unnavigable and meandering, it is ten feet deep in places and spreads out in a bewildering series of channels and swamps several miles wide. We used a pirogue that served as a resupply boat for the Japanese station. Propelled by poles, we crossed the river in shifts, and took advantage of the absolutely pure water to slake our parched throats. As we moved slowly through the river grass, we also passed through bouquets of fragrant pollen floating in the sunny air over the river. On a trip marked by every conceivable discomfort, the crossing was a moment of pure pleasure.

Without access to a pirogue, however, the crossing would be sheer hell, and this suggests one practical reason why the Pygmies never ventured into these forests. Even at its shallowest points, the river can take eight hours to cross on foot, and it is impassable much of the year. With ample game in more accessible forests, Fay says that Pygmy hunters have no need to risk a crossing.

Eventually we made our way to solid ground on the east bank of the river, and I felt a thrill. In essence this was a journey back in time. Entering the cool, wet forests, protected from the equatorial sun by the layers of canopy above us, we were entering the type of world not seen by most humans since the Pleistocene twelve thousand years ago, before our species invaded every nook and cranny on the planet. We started walking south, and from the first step we noticed that the animals were acting very strangely. We came upon a gang of red colobuses. Instead of fleeing, they just stared. The same was true of a gray-cheeked mangabey, and then a Peter's duiker.

At one point, we discovered leopard droppings containing black hair and some bone bits. The Pygmies claimed it was gorilla hair, though only DNA analysis eventually would tell for sure. Fay thought it was possible, since he has documented leopard attacks on gorillas. Through a pantomime, Samory, one of the trackers, demonstrated how these "perfect" predators kill the immensely strong apes. Impersonating a leopard, he showed how the cat will hide behind a rock and then, in a blinding flash, leap out and grab the gorilla's throat in its jaws. The Ndoki may be innocent of humans, but it is not a peaceable kingdom.

In fact, there is a civilization in these forests, even in the absence of humans. The area is latticed with trails, some as wide as boulevards, that have been cut and maintained by elephants. Ndokanda, a former elephant hunter turned tracker for Fay, says, "This is the elephant's city, and the leopard's and other animals' too." The grid of paths leads to the elephants' favorite spots: mineral licks and clearings, where they socialize with relatives and friends; baths, where they cover themselves with mud; knobby trees, where they rub the mud off, stripping their skin of ticks in the process; and trees such as the *Balanites wilsoniana* and *Autranella congoensis*, beloved by the big animals for their fruits.

Soon we had left behind the overhunted lands west of the Ndoki, where elephant trails were abandoned and overgrown. On the east side we saw fresh signs of elephants everywhere. A couple of times, they were so close we could even smell them. We did not, however, see or hear the giant mammals. Because of the vast territory they roam, and perhaps because of their ability to communicate with one another, they are the only creatures in this ecosystem that knew about humans. They know how dangerous we could be and stayed away from us, except for one day when they sent us a powerful warning by destroying our camp while we were gone, knocking over tents and smashing the makeshift wooden structures the Pygmy guides had erected so handily.

Still, Fay is heartened by all the signs that elephants are around. He points to a tree whose bark is still raw from being rubbed by an elephant. "That's what I like to see," he says. "There is nothing more depressing than seeing a tree with scars healed over because the elephants that used to rub against it are long gone."

Walking their paths and entering their clearings, all of which show evidence of recent use, is a little like walking into a house where the radio is playing and a pot is boiling on the stove but nobody is at home. The layout of the trails and clearings bespeaks an intelligence. (As in New York, the main routes run north and south and the smaller paths cut east and west.) The elephant community maintains this infrastructure through use, and—who knows?—perhaps through preventive maintenance chores assigned to different members of the group. There is a body of knowledge evident in the ways in which the elephants have altered the forest, and if some alien archaeologist stumbled into the Ndoki, he might interpret these earthworks as evidence of the accumulated knowledge of a sentient civilization. Why shouldn't we?

The elephants even have their own version of long-distance telephone calls, albeit more restricted than the version offered by AT&T. Suspecting that elephants employ very low-frequency sounds that are below the range of human hearing to communicate over long distances, Katy Payne devised an experiment to test this hypothesis on a group of elephants in East Africa. She broadcast low-frequency recordings of the estrus calls of two female elephants in heat. As she tells the story in her book *Silent Thunder*, two bull elephants picked up the signals and went rushing off, not towards two hot-blooded females, as they thought, but towards a Volkswagen Combi containing two young men and a couple of special loudspeakers. According to Payne, as the elephants thundered towards the researchers, the controller monitoring the action laconically asked the unsuspecting men in the Combi if they had any last wishes.

While it is tempting to impute semantic content to these long-distance messages, they may be richer in emotional content than in symbolism. Even if the messages broadcast are little more than "wish you were here," we can still imagine a rich community life in the "elephant city."

Many other animals benefit from elephant public-works projects. Their paths and clearings open up the forest for other big animals such as buffalo and ungulates like the bongo. The trail also certainly made walking easy for us. As we headed down one path, Joachine, one of the pygmy trackers, suddenly paused. The brush beside the trail erupted as a male gorilla charged. Just as suddenly, the big male stopped and just stared at us. This was an experience that was repeated time after time. Indeed, in encounters with fifteen groups of gorillas in the course of the expedition, we experienced only a couple of halfhearted charges. These charges were nothing like the terrifying charges many have experienced elsewhere in Africa, and after a while we began referring to the "pacifist gorillas of the Ndoki."

Like many of the other animals who watched our progress in blank-faced amazement, the gorillas seemingly did not know what to make of the alien invaders. We were clearly primates, bigger than chimps and tall by gorilla standards, but clearly not as well muscled as the puniest gorilla. In the abortive charges we could almost see the gorilla thinking, "Aha, invading primates, I'd better charge . . . wait, what are these guys? Why am I charging?"

Like other large animals, the gorillas have benefited from elephant land clearing and road building. Fay notes that when elephants clear the land, nutritious terrestrial herbaceous vegetation or THV springs up and is eaten by gorillas and a lot of big herbivores. Thus, the more elephants, the more gorillas. Fay says that the gorillas tend to favor marshy lowlands, while chimpanzees live farther from the water in the canopied forests. With both populations at very high levels, the Ndoki is one of the few places on earth where chimps and gorillas live close together. Researchers have seen both species in fig trees feeding on the energy-rich fruit at the same time.

The second day of the trip, we set off into what Fay called the "unknown." Our wafer-thin pretext for this mission was to find two clearings (or bais, as they are called) revealed by satellite imagery, and to field-test an early geographical positioning system (or GPS) that Fay had obtained so that he could better map this uncharted territory. The farther we got from villages and palm wine, the more the Pygmies came into their own. In the forest they are utterly self-reliant, creating cord from vines, cups from leaves, and bed mats from bark. At another research station fifty miles or so north, I remember a Pygmy glancing at a collection of forest seeds a scientist had gathered as part of the research for her dissertation. Without a moment's thought, the young man pointed to various seeds and said, "Gorillas eat that, chimps eat that, monkeys eat that," and so forth, casually confirming a couple of years of her research.

Even with fourteen years' experience in rain forests, Fay could still lose a trail, but Ndokanda, or any of the other Pygmies, can read the very faintest imprint with a glance. In fact, the only member of our group to lose a trail (naturally with me following him) was a Bantu porter who had married a Pygmy. Despite many years living and hunting with his Pygmy in-laws, the man had not acquired their preternatural ability to read the forest. On the third day, we had a dramatic demonstration of this astonishing Pygmy ability.

According to the satellite map, it looked as though we would have to cover almost twenty-five miles of dry land before reaching the next watershed. Unless we found a stream by dusk, we faced a waterless night after a full day's hike. Despite the offer of a good cigar as a bribe, Ndokanda set an uncharacteristically slow pace (the typical Pygmy pace through the forest is a blinding quickstep), so Fay decided to shame him by taking the lead. As we set off ahead, he remarked, "The one thing Pygmies can't stand is for a white guy to lead in the forest."

By afternoon I'm all sweated out and parched, but still we see no sign of water—or of the Pygmies straggling behind us. At one point, Fay spots a thick vine and says, "Aha!" He hacks off a section at just the right spot, and pure water spurts into his mouth. I grab his machete and hack away at the plant called *Sissus danclydgia*, but manage to taste only a few remaining drops.

As the sun sinks and it appears that we will spend a dry and desperate night, we finally hit sandy soil—a good sign. Soon we find elephant footprints filled with water. It looks pure, and I drink greedily. Fay's hand is so tired from hours of hacking with the machete that he cannot open the water bottle I have just filled.

Just before dark, Ndokanda comes motoring by us. Not bothering to stop, he yells at Fay in Sango, his Pygmy language, "You fool, I know this place. Right ahead there is plenty of water." Ndokanda is right, of course, and we are left openmouthed, wondering what enabled him to recall this tiny part of a vast forest from a brief foray with Fay years earlier.

After a couple of days exploring the area, we set off deeper into the forest. Trips with Fay are a bit like a treasure hunt—if you consider the half-eaten remains of a fruit discarded by a gorilla to be treasure. Fay enthusiastically sampled these fruits, and I tried the juicy kernels of a *Myrianthus arboreus*. This may have been the moment I picked up a mysterious gastrointestinal disorder that took a couple of years of consultations with tropical medicine experts and rain forest healers to cure. Or perhaps I got the bug from the partially eaten *Treculia africanus*, a fruit favored by Pygmies, gorillas, and chimps, and which, I discovered, tastes a little like peanuts.

During these walks we would occasionally stop and ask the Pygmies to call duikers. They do this by holding the bridges of their noses and making a loud braying sound in imitation of the sounds made by these small deerlike animals when giving birth. Other duikers come running when they hear the sound, which makes hunting easy for the Pygmies. Hunting, however, was not our purpose. Among the other creatures attracted by the braying sound were chimpanzees who see this as an opportunity to do some hunting of their own and catch a duiker at a vulnerable moment.

Stopping intermittently to make the calls, we attracted several unusual animals, including the rare yellow-backed duiker, an animal whose dull golden patch on its back supposedly gave rise to the myth of the Golden Fleece. Then, pausing for a rest, we hit pay dirt. Fay, Karen Lotz, and I were ahead of the rest of the group of Pygmies along with Ndokanda. Ndokanda hunkered down and made the call. This time a group of large animals came crashing towards us, and for a moment I felt the shiver of being hunted.

That feeling vanished as soon as a very large band of chimpanzees appeared out of the brush and saw us. They stopped dead in their tracks. Bloodlust gave way to astonishment. It was quite clear that they were seeing something they had never seen before. They began stamping their feet, shaking their arms, calling to one another, and, occasionally, throwing branches at us. Little ones ventured bravely towards us, only to be pulled back by their protective moms. In the branches above us, a very old chimp with completely white hair gazed down on us slack-jawed with amazement. I wonder whether the other chimps would later turn to him for an explanation of these otherworldly visitors.

As many as twenty-five animals screamed at us from all sides as we maintained a studied, casual stance, minimizing jerky movements. Each time we made a move, a new round of calls erupted among the chimps, but they never showed signs of fleeing, and they never attacked. Wild chimps do not react this way to humans in any other part of the African rain forest. For more than two hours, the mesmerized chimps hovered around us, drawing to within a few arm's lengths.

Later Fay called this the signal wildlife experience of his fourteen years in Africa. For me it was the experience of a lifetime. For the chimps surrounding us, seeing humans amounted to an ape version of *Close Encounters of the Third Kind*. The ruckus the apes raised began with threats and distress calls, but some of the apes seemed to let out the hoots that chimps use to greet one another. I would like to think that at least some of these chimps were welcoming us ape-like aliens into their forest.

With every encounter, I became more convinced that this forest, empty of humans, is not empty of intelligence of various sorts. There is the accumulated knowledge of the elephant civilization that gives this forest its distinctive flavor. There are the chimp bands whose members scheme and forge alliances for their own advancement, who make and use tools to get food, and who cooperate with each other when hunting or in conflict with other bands. There are the gorilla families dominated by silverbacks who must be alert to treachery in their harems and plotting by ambitious young males. There is some measure of awareness in the leopards who must learn a host of different skills in stalking and killing in their never-ending search for prey.

While none of these animals have the extraordinary powers of concentration and symbolic expression humans developed in the course of evolution, they may have other mental gifts that humans lack or have lost over time. Chimps will band together at night and then go off in several different directions during the day, only to meet again before settling down for the evening. How do they communicate the plan of the day and divide responsibilities if that is what they do? Do these and other animals have nonsymbolic ways of planting images in each other's consciousness? Apart from propositional abilities, is there some entirely different form of sentience binding the creatures of this forest? It is possible, but we do not need to invoke the paranormal to appreciate that this ecosystem without humans still hums with intelligence. And it is comforting to know that a tree falling in the forest is reported even if humans are not there to notice the fall.

Still, the presence of intelligence in the forest does not necessarily mean it is a friendly place. Sometime towards the end of the expedition, one of the many microbes lying in wait for a new organism to invade found its way to me. In no position to diagnose the ailment, I described it in terms of its symptoms as the "heart attack/broken rib/deadly tick" disease (not to be confused with the nearly incurable gastrointestinal illness I also picked up in the forest). Later, Dr. Kevin Cahill, a well-known specialist in tropical medicine, guessed that it was probably dengue fever. Whatever it was, its appearance was unwelcome and inopportune in the middle of the Ndoki. Weakened and feeling like every bone in my body had been broken, I decided that it would be better to suffer through a long, arduous walk to the dubious comforts of Bomassa than to take my time and risk further decline in the middle of the Ndoki.

Thus, we hiked out nearly thirty miles in one day. Mike takes a casual approach to equipment, and the shortcomings of this attitude became apparent towards evening as we retraced our path to the Djeke River camp. There we left the Pygmies, who wisely decided to wait for morning light, and continued by flashlight as darkness fell on the forest like an anvil. Unfortunately, all of our flashlight batteries quickly failed as well. Karen produced one of those tiny pen flashlights used for map reading, and our ludicrous procession of the nearly blind leading the halt continued as Mike tried to find signs of the trail with the feeble light penetrating a few inches of gloom in front of him. After an hour or so of this, we groped our way onto the more well-trodden trail leading into Bomassa, and finally, with every one of my joints in open rebellion, we staggered into camp a few hours later. Now I knew the meaning of the words "very, very difficult."

I went to the forest in 1992 with the ambition of bringing its wonders to light and perhaps mobilizing world opinion to protect this rarity. Since then the Congo has had coups and uprisings, but thanks in part to the efforts of the Wildlife Conservation Society and other conservation groups, and the massive clout of the World Bank, the region has remained largely inviolate. It is now called Ndoki-Nouabale National Park, and if people abide by the agreement, the core area will remain forever off limits to everyone but certified scientific expeditions operating under strict controls. This is a big if.

A small area, including parts of the region I walked in 1992, has been opened to low-impact ecotourism. To facilitate entry to the forest, a road has been built from Bomassa to the banks of the Ndoki. A trip that once consumed much of two days can now be accomplished in less than an hour by Jeep. Continuing turmoil in the region has kept the number of visitors to a few hardy souls a month, but making any area accessible in Africa tends to have catastrophic consequences for wildlife. Mike Fay's own studies have shown that the greatest amount of poaching is almost always within a day's walk of the nearest road.

When I was in the Ndoki, Fay and I had a number of discussions about the dangers of demythologizing the Ndoki for the Pygmies, who might someday be tempted by its readily available bush meat of all sorts. Subsequent to my trip, I did hear that at least one band of Pygmies had been caught hunting in the Ndoki. The poaching raid did not involve any of the Pygmies we had hired, but to my dismay, I learned that the group had been led there by an American who had lived with the Pygmies for a number of years and who knew Fay. This particular group, based in the Central African Republic, had been having trouble finding game, and the American had brought them into the Ndoki so they could hunt. The best protection for the Ndoki had been the fears of those who might otherwise be tempted to hunt there. Once those fears are proven to be unfounded, no amount of Western aid will be able to stop poachers from entering the forests.

There have been other developments in these forests since my trip in 1992. When I went to the Ndoki, only a few scientists had spent much time there, and then mostly on its periphery. At that point, there was no evidence that any humans had ever established themselves in the region. Since then subsequent expeditions have uncovered ancient palm nuts in riverbeds. Because only humans would bring the palm nuts to the river, these fossils do provide evidence that the forests were inhabited in the distant past. Carbon dating of these nuts suggests that human numbers in the region peaked about seventeen hundred years ago and then quickly vanished by A.D. 700.

Other research in contiguous forests in Gabon suggests that human numbers crashed throughout central Africa at that time. Massive deforestation either preceded or accompanied this calamity, but the disappearance of population pressures subsequently gave the forests time to recover. It is still unclear what caused this ecological and human catastrophe, but the dates are suspiciously close to a worldwide episode of climate change that saw the beginning of a three-hundred-year drought in Mesoamerica (at about the time that the Mayans began their decline), as well as the disruption of rainfall patterns in Europe and Asia.

The finding made me think again of the ominous pattern of drying over the past twenty-five years that has brought the Ndoki to the lower threshold of rainfall necessary to support a wet tropical forest. Is the region doomed no matter what the nations of the world try to do to save it? In what ways have we contributed to these ominous climate changes through deforestation elsewhere in Africa and the burning of fossil fuels? It seems that animals in the most remote places on earth are still captive to human activities even if they have never seen a human being.

I find it unutterably sad that we are coming to understand that the world is quickened with intelligence even as we extinguish the forests and the denizens whose various minds give it life. The arrogance of humanity has been our assumption that we are the only sentient beings in a world otherwise empty of consciousness and interior life. Drunk with our own intellectual power, we have overwhelmed the rest of nature with our tinkerings, crowding out and killing off other species at a rate not equaled since the calamity that wiped out the dinosaurs 65 million years ago. Nature will ultimately recover, of course, but we may pay dearly for the hubris of believing that we can run the biosphere better than nature, which has had billions of years to let the patient workings of evolution optimize various ecosystems for stability and resilience.

What is intelligence, anyway? If life is about perpetuation of a species, and intelligence is meant to serve that perpetuation, then we can't hold a candle to pea-brained sea turtles who predated us and survived the asteroid impact that killed off the dinosaurs. During our brief tenure of 200,000 years on this planet, we have pushed the biosphere's life-support system to the limit and played dice with the ozone layer and climate, two necessary preconditions for a habitable world and sustenance. How intelligent is that? Does anyone seriously think that we can continue use our gifts for invention and communication for another few thousand years without bringing upon ourselves a man-made apocalypse—much less the 165 million years that turtles have found a way to survive.

As we are proving, without some controls to represent the long-term interests of the biosphere, our brand of intelligence is dangerous. Perhaps it has come and gone several times in different species in the past. The unfettered application of propositional abilities does not seem to be a prescription for long-term evolutionary success. Once minds break free of religious, cultural, and physical controls, they burn hot and fast, consuming and altering everything around them. Perhaps this is why higher mental abilities, though present in other creatures, are more limited and circumscribed.

A good deal of the blood that goes to the brain in humans goes to the muscles in our closest relatives, the chimps and bonobos. The brain is literally subordinated to the demands of the body. It's a trade-off that is the rule in nature rather than the exception. One doesn't have to impute a plan to evolution to see an implied judgment that long-term success comes to those creatures who fit the needs of larger ecosystems rather than those who presume to alter ecosystems to their needs.

Even if shackled, however, some intelligence illuminates the lives of countless creatures. When we deal with captive animals and when we encounter them in the wild, we can see it peeking out in a flash of brilliance here and there as animals draw on abilities that help them secure food and prosper within their communities in order to manipulate, deceive, beguile, and otherwise deal with the humans they encounter. Every so often, they do something extraordinary, and we gain some insight into where some of our abilities might have originated. Perhaps more important, we can then understand how it might be fun to be an orangutan or a parrot. They and many other creatures enjoy a life rich in emotions and physical prowess but still are gifted with the ability to appreciate life from a distance, even if its horizons are more constrained than those of the heady, perilous view from Olympus that is our blessing and curse.

SELECTED BIBLIOGRAPHY

Boesch, Christophe. 1992. "New elements of a theory of mind in wild chimpanzees." *Behavioral and Brain Sciences.*

Byrne, Richard W. 1995. *The Thinking Ape.* Oxford: Oxford University Press.

———, and Andrew Whiten. 1988. *Machiavellian Intelligence: Social Expertise and the Evolution of Intellect in Monkeys, Apes, and Humans.* Oxford: Oxford University Press.

Chadwick, Douglas H. 1994. *The Fate of the Elephant.* San Francisco: Sierra Club Books.

Darwin, Charles. 1872. *The Expression of the Emotions in Man and Animals.* Reprint, Chicago and London: The University of Chicago Press, 1965.

Dennett, Daniel C. 1991. *Consciousness Explained.* Boston, Mass.: Little Brown.

de Waal, Frans. 1982. *Chimpanzee Politics: Power and Sex Among the Apes.* New York: Harper and Row.

———. 1996. *Good Natured: The Origins of Right and Wrong in Humans and Other Animals.* Cambridge, Mass.: Harvard University Press.

Durrell, Gerald Malcom. 1979. *My Family and Other Animals.* New York: Viking.

Frisch, Karl von. 1967. *The Dance Language and Orientation of Bees.* Cambridge, Mass.: Harvard University Press.

Gallup, G. 1982. "Self-awareness and the emergence of mind in primates." *American Journal of Primatology.*

Griffin, Donald. 1984. *Animal Thinking.* Cambridge, Mass.: Harvard University Press.

———. 1998. "From cognition to consciousness." *Animal Cognition.*

Helfer, Ralph. 1997. *Modoc: The True Story of the Greatest Elephant Who Ever Lived.* New York: HarperCollins.

Holldobbler, Bert, and Edward O. Wilson. 1990. *The Ants.* Cambridge, Mass.: Harvard University Press.

Leakey, Richard, and Roger Lewin. 1993. *Origins Reconsidered: In Search of What Makes Us Human.* New York: Anchor.

Linden, Eugene. 1974. *Apes, Men and Language.* New York: Penguin.

———. 1986. *Silent Parners: The Legacy of the Ape Language Experiments.* New York: Times Books.

———. 1992. "Apes and humans." *National Geographic.*

———. 1994. "Can animals think?" *Time.*

Lopez, Barry Holstun. 1978. *Of Wolves and Men.* New York: Scribner's.

Lorenz, Konrad. 1966. *On Aggression.* New York: Harcourt Brace.

Marler, P., and C. Evans. 1996. "Bird calls: Just emotional displays or something more?" *Ibis.*

McGrew, William. 1992. *Chimpanzee Material Culture: Implications for Human Evolution.* Cambridge, U.K.: Cambridge University Press.

Midgley, Mary. 1978. *Beast and Man: The Roots of Human Nature.* Ithaca, N.Y.: Cornell University Press.

Nagel, Thomas. 1974. "What is it like to be a bat?" *Philosophical Review* 83.

Patterson, Francine, and Eugene Linden. 1981. *The Education of Koko.* New York: Holt, Rinehart & Winston.

Payne, Katy. 1998. *Silent Thunder: In the Presence of Elephants.* New York: Simon and Schuster.

Peterson, Dale. 1989. *The Deluge and the Ark.* Boston, Mass.: Houghton Mifflin.

Premack, David, and G. Woodruff. 1978. "Does the chimpanzee have a theory of mind?" *Behavior and Brain Science.*

Pryor, Karen. 1995. *On Behavior.* North Bend, Wash.: Sunshine.

———, Richard Haag, and Joseph O'Reilly. 1969. "The creative porpoise: Training for novel behavior." *Journal of the Experimental Analysis of Behavior.*

Ristau, Carolyn A., ed. 1991. *Cognitive Ethology: The Minds of Other Animals* (Essays in Honor of Donald Griffin). New Jersey: Erlbaum.

Savage-Rumbaugh, Sue, and Roger Lewin. 1994. *Kanzi: The Ape at the Brink of the Human Mind.* New York: John Wiley & Sons.

Sebeok, Thomas A., and R. Rosenthal, 1981. "The Clever Hans phenomenon: Communication with horses, whales, apes, and people." *Annals of the New York Academy of Sciences.*

Strum, Shirley C., and George B. Schaller. 1990. *Almost Human: A Journey into the World of Baboons.* New York: W.W. Norton.

Tattersall, Ian. 1998. *Becoming Human.* New York: Harcourt Brace.

Whiten, Andrew, ed. 1991. *Natural Theories of Mind: Evolution, Development and Simulation of Everyday Mindreading.* Oxford: Blackwell.

Wrangham, Richard W., ed., et al. 1994. *Chimpanzee Cultures.* Cambridge, Mass.: Harvard University Press.

INDEX